THREE ANCIENT GEOGRAPHICAL TREATISES IN TRANSLATION

This volume is a translation and commentary on the works of three geographers from Greco-Roman antiquity: Hanno of Carthage, from around 500 BC; the author of the *Periodos Dedicated to King Nikomedes*, from the last half of the second century BC; and Avienus, from the fourth century AD.

The modern translations of texts in this book represent 1,000 years of Greco-Roman geographical scholarship, and thus provide an overview of the discipline from its beginnings to late antiquity. Readers will learn about the development of Greek geography, and the earliest adventures outside the Mediterranean into the Atlantic, as far south as the tropics and north toward the Arctic. These explorations make for fascinating stories about early human endeavors into an unknown world.

Three Ancient Geographical Treatises in Translation offers specialists new information about Greek exploration and a modern translation of significant ancient texts, while non-specialist scholars and undergraduate students with an interest in Greco-Roman literature and ancient geography will also find the volume useful and accessible.

Duane W. Roller is Professor Emeritus of Classics at the Ohio State University, USA, having received a Ph.D in Classical Archaeology from Harvard University, USA, and is the author of *Cleopatra: A Biography*, *Cleopatra's Daughter and Other Royal Women of the Augustan Era*, and *Empire of the Black Sea: The Rise and Fall of the Mithridatic World*.

ROUTLEDGE CLASSICAL TRANSLATIONS

Routledge Classical Translations provides scholars and students with accurate, modern translations of key texts that illuminate distinctive aspects of the classical world and come from a range of periods, from early Greece to the Byzantine empire. Volumes include thematic groupings of texts, texts from important authors as well as texts from the Byzantine period that are relevant for the study of the classical world but which remain inaccessible. Each volume has accompanying notes and commentary that provide a solid framework for deeper understanding of the material. As well as providing translations of significant texts, the series makes available material that is untranslated into English or difficult to access, and places these texts within new contexts to open-up areas of study and support research.

Titles include:

PLUTARCH'S THREE TREATISES ON ANIMALS
A Translation with Introductions and Commentary
Stephen T. Newmyer

MEGASTHENES' *INDICA*
A New Translation of the Fragments with Commentary
Richard Stoneman

THREE ANCIENT GEOGRAPHICAL TREATISES IN TRANSLATION
Hanno, the *King Nikomedes Periodos*, and Avienus
Duane W. Roller

PYTHEAS OF MASSALIA
Texts, Translation, and Commentary
Lionel Scott

For more information about this series, please visit: https://www.routledge.com/Routledge-Classical-Translations/book-series/CLTRA

THREE ANCIENT GEOGRAPHICAL TREATISES IN TRANSLATION

Hanno, the *King Nikomedes Periodos*, and Avienus

Duane W. Roller

Routledge
Taylor & Francis Group

LONDON AND NEW YORK

First published 2022
by Routledge
2 Park Square, Milton Park, Abingdon, Oxon OX14 4RN

and by Routledge
605 Third Avenue, New York, NY 10158

Routledge is an imprint of the Taylor & Francis Group, an informa business

British Library Cataloguing in Publication Data
A catalogue record for this book is available from the British Library

Library of Congress Cataloging-in-Publication Data
Names: Roller, Duane W., author. | Hanno. Periplus. English. | Pseudo-
Scymnus. Orbis descriptio. English. | Avienus, Rufius Festus. Ora maritima.
English.
Title: Three ancient geographical treatises in translation : Hanno, the King
Nikomedes Periplous, and Avienus / Duane W. Roller.
Description: Abingdon, Oxon ; New York, NY : Routledge, 2022. |
Series: Routledge classical translations | Includes bibliographical references
and index. |
Identifiers: LCCN 2021018925 (print) | LCCN 2021018926 (ebook) |
ISBN 9780367462543 (hardback) | ISBN 9781032112916 (paperback) |
ISBN 9781003030379 (ebook)
Subjects: LCSH: Classical geography. | Hanno. Periplus. | Pseudo-Scymnus.
Orbis descriptio. | Avienus, Rufius Festus. Ora maritima.
Classification: LCC G84 .R649 2022 (print) | LCC G84 (ebook) |
DDC 913--dc23
LC record available at https://lccn.loc.gov/2021018925
LC ebook record available at https://lccn.loc.gov/2021018926

ISBN: 978-0-367-46254-3 (hbk)
ISBN: 978-1-032-11291-6 (pbk)
ISBN: 978-1-003-03037-9 (ebk)

DOI: 10.4324/9781003030379

Typeset in Times New Roman
by Taylor & Francis Books

CONTENTS

MAPS

ABBREVIATIONS

ANRW	*Aufstieg und Niedergang der römischen Welt*
AntAfr	*Antiquités africaines*
BAR	*British Archaeological Reports*
BCTH	*Bulletin archélogique du Comité des Travaux Historiques et Scientifiques*
BNJ	*Brill's New Jacoby*
BNP	*Brill's New Pauly*
BSA	*Annual of the British School at Athens*
CAH	*Cambridge Ancient History*
CIL	*Corpus inscriptionum latinarum*
CR	*The Classical Review*
EANS	*The Encyclopedia of Ancient Natural Scientists* (ed. Paul T. Keyser and Georgia L. Irby-Massie, London 2008)
FGrHist	*Die Fragmente der Griechischen Historiker*
G&R	*Greece and Rome*
GGM	K. Müller, *Geographi graeci minores* (Paris 1855–61)
GJ	*Geographical Journal*
HG	*Hellenistica Groninga*
JRAS	*Journal of the Royal African Society*
JRS	*Journal of Roman Studies*
MCV	*Mélanges de la Casa de Velázquez*
MUSJ	*Mélanges de l'Université Saint-Joseph*
OGIS	*Orientis graeci inscriptiones selectae*
OrSue	*Orientalia Suecana*
PECS	*Princeton Encyclopedia of Classical Sites*
PP	*La parola del passato*
QUCC	*Quaderni Urbinati di Cultura Classica*
RA	*Revue archéologique*
RAN	*Revue archéologique de Narbonnaise*
RE	*Paulys Real-Encyclopädie der Classischen Altertumswissenschaft* (Pauly-Wissowa)
RFIC	*Rivista di filologia e di istruzione classica*

RPA	*Revista portuguesa de Arqueologia*
RSA	*Rivista storica dell'Antichità*
RStudFen	*Rivista di studi fenici*
SIG	*Sylloge inscriptionum graecarum*
TAPA	*Transactions and Proceedings of the American Philological Association*

PREFACE

The extant literary history of Greek geographical writing extends over nearly 1,000 years, from the summary of the West African expedition of Hanno of Carthage, dating from the fifth century BC, to the Greek material in the *Ora Maritima* of Avienus, written in the fourth century AD. Geography was a common literary genre in the Greco-Roman world—over 200 authors are known by name—but only a handful of texts survive. The number of major treatises is limited (most notably Strabo, Pomponius Mela, Pliny, and Ptolemy), but there are a larger number of short works extant in whole or in part. In late antiquity about 20 of them were collected in a corpus known familiarly today as the *Minor Greek Geographers*, in which two of the texts in the present volume, Hanno and the King Nikomedes *Periodos*, appear.

Hanno's report, or the extant summary of it, is perhaps the earliest existing treatise in the ancient geographical tradition and a rare insight into the world of Carthaginian exploration, whose results became tangled into the Greco-Roman geographical experience. It describes in summary form a voyage from Carthage out into the Atlantic and down the West African coast as far as the tropics. The anonymous text dedicated to a King Nikomedes, whose author has been known conventionally but unhelpfully since the seventeenth century as Pseudo-Skymnos, dating from the late second century BC, stands within the flourishing Hellenistic world of rich and thorough geographical writing, and, insofar as it is preserved, presents an account of the coasts of Europe and parts of Asia.

From another manuscript tradition is the poem of Rufus Festus Avienus, the *Ora Maritima*, an itinerary from Brittany south along the Atlantic coast and into the Mediterranean as far as Massalia (modern Marseille): the continuation of the account to the east has been lost. Although in Latin, and also one of the latest geographical treatises to survive from classical antiquity, it is fully within the tradition of Greek geographical scholarship, and also relies on Carthaginian material.

Common to all three authors is the concept of the *periplous*, the coastal sailing guide that formed the original Greek basis of geographical interest. The Greek intimacy with seafaring, and the need to understand distances, directions, and seaborne perils, meant that from earliest times accounts of routes from place to place were an essential part of the Greek experience. Presumably these were originally oral reports, such as the instructions Kalypso gave to Odysseus, but they eventually evolved into a literary genre. Actual *periploi* exist, but the format pervades all Greco-Roman geographical writing, including the major extant texts such as Strabo and Pliny. Even though the titular heading of the treatise dedicated to King Nikomedes calls it a "circuit of the earth" (*periodos ges*), its format, with its emphasis on the coasts, is derived from a *periplous*.

In the chapters that follow, each of the three authors is treated separately, after a preliminary discussion of the nature of the geographical tradition that they represent and its role in the history of ancient geography and exploration. In addition to the translation of the texts of Hanno, the King Nikomedes *Periodos*, and Avienus, and the related commentaries, there is an examination of the manuscript tradition of each author and their position within the genre of ancient geographical writing. Covering as they do hundreds of years of Greek scholarship (even if, in one case, represented in the Latin language), the three texts allow understanding of the evolution of the genre throughout Greco-Roman antiquity.

Even though the three authors are discussed separately, and extend over nearly 1,000 years of Mediterranean history, using two languages (three if the Punic original of Hanno's report is considered), there is a remarkable overlap between them. All three consider in their own way the region of the Pillars (Columns) of Herakles (Hercules) and the surrounding territory, especially the ancient Phoenician trading post of Gadeira or Gadir (modern Cádiz). Hanno and Avienus both examine the world beyond the Pillars—the Atlantic coasts—and the King Nikomedes *Periodos* and Avienus describe the European coast of the western Mediterranean. Each, of course, has its own outlook and biases, with differing and contradictory topographical conclusions. But all demonstrate how the Pillars of Herakles are the nexus of ancient geographical thought.

The present translator and commentator has drawn on his experience in the study of ancient geography, including a number of critical analyses of the discipline as well as his translations of and commentaries on major geographical texts, such as those of Megasthenes, Eratosthenes, Strabo, and Pliny the Elder. He would like to thank Georgia L. Irby, Carolina López-Ruiz, Letitia K. Roller, Esther Rodríguez Gonzáles, Lisbet Thoresen, the Emeritus Academy of the Ohio State University, and Amy Davis-Poynter and many others at Routledge. As usual, the writing was done in his study in Santa Fe.

INTRODUCTION

The Periplous

When Odysseus grew tired of the delights of Kalypso's island of Ogygie and wanted to return home, he faced two problems. He had no transportation, and he had no idea where home was. The first matter was solved when he built—with the goddess' help—a boat for himself. Homer's description of its construction is an important early Greek account of the craft.[1] But the other issue was more serious: Odysseus did not know where he was, and, more importantly, did not know how to reach Ithaka. Again Kalypso provided him with the advice he needed:

> He looked upon the Pleiades, late-setting Boötes, and the Bear, which is called the Chariot and which is always rotating where it is and watches Orion, having no share in the baths of Okeanos. Kalypso, the shining goddess, had ordered him to keep it on his left hand as he sailed the sea.

This is the earliest example of sailing instructions in Greek literature, using the constellations as points of reference. Although Odysseus might not have been totally aware of it, they also revealed—with the Bear constantly on his left—that Ithaka was somewhere to the east of Ogygie. Regardless of the details, and how the Homeric account relates to actual events of the early Greek world, Kalypso's information established the nature of Greek geographical enquiry, which originated in laying out sailing routes in the Mediterranean. Her brief statement is primitive but contains crucial information, and although it would take Odysseus quite some time to make it back to Ithaka, in theory she gave him all the data that he needed.

Seamanship in Greek waters had existed since the Neolithic period, and ships were depicted in art by the third millenium BC.[2] Any sailor, however brief the journey, needed some knowledge of the route to be taken. This information would have been transmitted orally for hundreds of years—through conversations in waterfront tavernas—but from the time of Homer such material began to be written down.

DOI: 10.4324/9781003030379-1

The literary genre that was the intellectual descendant of Kalypso's instructions was the *periplous*. The original meaning of the word is "a sailing around," first documented when the fleet of Dareios I of Persia refused to sail around Mt. Athos because of the known dangers of the route.[3] Exactly when the term evolved into the name of a literary genre is uncertain: Skylax of Karyanda, active in the late sixth century BC, allegedly wrote a *Periplous Outside the Pillars of Herakles*,[4] but the title is only documented in the tenth century AD and there is no assurance that Skylax himself used the word. Nevertheless he was an important pioneer in Greek seamanship, sent by Dareios to make a seaborne reconnaissance of the eastern portions of the Persian empire.[5] If he did write the *Periplous Outside the Pillars of Herakles*, it does not fit with other known details of his life, and means that he went far afield at a remarkably early date, but his career is so vaguely known that it cannot be said with certainty whether or not he originated the concept of the *periplous*.

Yet there is little doubt that the idea of a literary *periplous* was well enough established by the fifth century BC that the translator of the summary report of the expedition of Hanno could use it as the titular opening of the text: "The *Periplous* of Hanno, King of the Carthaginians."[6] From that time the genre became an integral part of Greek geographical writing. Even if an account is not specifically presented as a *periplous*, the form is buried in much of Greek literature that includes geographical data. Telltale markers of a hidden *periplous* appear frequently, such as Thucydides' comment, in the context of events leading up to the Peloponnesian War, that Epidamnos is on the right as one sails into the Ionian Gulf.[7]

Periploi were an essential component in the development of ancient geography. In earliest times, as people began to travel in the eastern Mediterranean, going from one island to another or from one coastal city to another, data began to be collected about the experiences of sea voyages. Even the shortest journey required knowledge of winds, directions, and geographical features that could serve as navigational reference points.[8] A basic understanding of celestial phenomena, as Odysseus knew, was also a necessity. He also realized that sailing into unknown lands meant encounters with different peoples, who could be hostile. Although the *Odyssey* expresses Odysseus' experiences in a mythical form, groups such as the Laistrygonians or Lotus Eaters, or the world of Circe, would be genuine issues, although perhaps not in the same way that their memories were transmitted by Homer. Navigational hazards, many of which still exist in the Mediterranean, were also a problem. There were the whirlpool of Charybdis, the rocks of Skylla, the shoals of the Sirens, or the moving Symplegades Islands. As people sailed more, especially after the end of the Bronze Age, a body of received wisdom began to develop, oral at first and probably highly fantasized—as the Homeric poems show—but no less informative or dangerous. Eventually such material came to be written down, and the *periplous* as a literary genre began to develop.[9]

The texts examined in this volume preserve vestiges of the emergent *periplous* format. The extant Greek translation of the expedition of Hanno, from the fifth century BC, is probably the earliest unquestioned use of the word in Greek to describe an account of a coastal voyage. The *Ora Maritima* of Avienus, although written in its present form hundreds of years later and not in Greek, nevertheless is based on *periploi* of a much earlier era, probably of Massalian or Phokaian origin and roughly contemporary with the cruise of Hanno. However condensed and fragmentary these reports are, they demonstrate that as a literary form the genre was well established by the fifth century BC. Many later *periploi* exist, in whole or in part, or within other texts. The genre especially flourished in and after the time of Alexander the Great, when new worlds were explored both east and west of the Greek heartland: examples include the reports of Nearchos and Onesikritos, who led the expedition from the Indos River back to the Persian Gulf under the orders of Alexander, Androsthenes of Thasos, who explored the Persian Gulf at roughly the same time, and Patrokles, who travelled around the Caspian.[10] Eventually, in the late second century BC, the second text in this volume was produced, dedicated to a certain King Nikomedes and a reflection of the state of geographical knowledge in the late Hellenistic period, two centuries after the time of Alexander the Great and before the final effects of Roman expansionism. Although not a true *periplous* and not titled as such, it is beholden to the format, using the coasts as its basic frame of reference, yet with more intrusions inland than one would expect in an actual *periplous*.

Geographical theory

During the years that changing Greek geographical horizons led to the creation of sailing manuals, there was a parallel development of geographical theory. The seamen, merchants, and traders for whom the *periploi* were designed had little interest in such concerns, but the construction of a theoretical basis for geographical knowledge impacted the overall conceptions of the earth and its surface and thus were an essential component of geography.

Thales of Miletos, the originator of the Greek intellectual movement, active in the late seventh and early sixth centuries BC, suggested that the earth rests on water, floating like a piece of wood, or even like a ship.[11] This shows that even by Thales' day seamanship was so embedded in the Greek consciousness that nautical analogies were having an impact on developing geographical theory. But the most revolutionary idea was that the earth was a sphere. This point of view was Pythagorean in origin and was perhaps first published by Parmenides of Elea in the fifth century BC.[12] Sailors had long realized that the surface of the sea was curved, since ships rose above or sank below the horizon. There was also an understanding that celestial phenomena changed as one went north or south. But the Pythagorean belief in the spherical earth was more harmonic than empirical. In fact, it was so counter-intuitive that it was not quickly accepted: even as late as the third century BC it was not taken for granted.[13]

Parmenides was also the first to theorize about terrestrial zones: the concept of bands of similar climate that went around the earth at the same latitude. At the far north there was a frozen zone and at the equator a burned zone, with the inhabited temperate zone between them.[14] This was the beginning of climate theory, a crucial part of understanding the earth as a whole.

With the belief that the earth was a sphere, it was natural to wonder about its size. The earliest known figure for this is 400,000 stadia, probably due to Eudoxos of Knidos in the early fourth century BC.[15] This and other figures proposed in the fourth and early third centuries BC were not based on any attempt at measurement, but were educated guesswork. It was only with Eratosthenes of Kyrene, in the second half of the third century BC, that there was an accurate mathematical determination of the circumference of the earth, at 252,000 stadia. His calculations were published in his treatise *On the Measurement of the Earth*, which survives today in a handful of fragments.[16] It was also Eratosthenes who invented the discipline of geography and coined the word *geographia* to describe his efforts, outlining his conclusions in another treatise, his *Geographia*, which exists in about 150 fragments. He positioned the inhabited portions of the earth on its sphere, and located several hundred known places, thereby realizing that they were spread across only a small part of the terrestrial surface and totally within the northern hemisphere. In his identification of the portion of the earth that was known to be inhabited–the Mediterranean basin and its surroundings–Eratosthenes popularized a recent term, *oikoumene* (literally "inhabited region"), that may have been invented by Aristotle a century or more previously.[17]

Notwithstanding further refinements, the work of Eratosthenes established the structure of ancient geographical theory. The size and shape of the world were known, and there was an understanding of some of its climatic phenomena. Moreover, the nature and location of the *oikoumene* had been determined. To be sure, measurements would be handled with greater precision, and more and more places would be positioned, to a maximum of about 8,000 in antiquity.[18] But by the late Hellenistic period, ancient geography was a recognized discipline and in the mainstream of Greek intellectualism.

The sources

The Greek translation of Hanno's report is earliest preserved in a manuscript in Heidelberg, Palatinus Graecus 398, which dates to the ninth century and is a collection of various brief texts, largely geographical. These may have been grouped together by Markianos of Herakleia Pontika, who lived sometime between the second and sixth centuries AD and was a geographical scholar who remains elusive today, although some fragments of his writings survive.[19] Markianos seems to have collated a corpus of geographical writings, including his own. In whatever way it was distributed, by the ninth century it had split into several manuscript traditions, of which Palatinus Graecus 398 is the earliest extant.

In the mid-nineteenth century Markianos' collection came to be called the *Minor Greek Geographers*, published by Carl Müller as the *Geographi Graeci Minores* (Paris 1855–61). There has been no complete publication of the collection since then, although individual texts had appeared previously, including that of Hanno in 1533, the earliest edition of that author. There have been a number of recent editions of his text, of varying quality.[20]

The *Periodos Dedicated to King Nikomedes*, whose anonymous author is erroneously but commonly known as Pseudo-Skymnos, is also part of the corpus of *Minor Greek Geographers*. It is not preserved in Palatinus graecus 398, but in Codex Parisinus graecus supplementi 443, a poor-quality manuscript of the thirteenth century AD, representing a tradition that diverged from Markianos' collection sometime before the ninth century. Its end is lost, and since the *Periodos* is the last item, the second half of it is no longer extant, which would have included the name of the author at the end of the text. A few additional fragments are preserved in another manuscript, Codex Vatopedinus 655, of the fourteenth century, and located in the Vatopedi monastery on Mt. Athos. These fragments of the *Periodos* are buried in an anonymous work titled the *Periplous of the Euxine Sea*.[21] Although some copies of material in Codex Parisinus graecus supplementi 443 were made between the thirteenth and fifteenth centuries, the manuscript itself was lost and did not resurface until 1837, and then was published as part of Müller's *Geographi graeci minores*. The current translation is based on the Budé edition of Didier Marcotte with the additional material from the *Periplous of the Euxine Sea* as collated by Aubrey Diller.[22]

The *Ora maritima* of Avienus is from a totally different manuscript tradition, which is actually completely lost. All editions of the text depend on a 1488 publication by Victor Pisanus. There are several modern editions, including that by Dietrich Stichtenoth (1968), J. P. Murphy (1977), and Luca Antonelli (1998), which form the basis of the present translation.

Transliteration

Transliteration of toponyms preserved in Greek is always a problem, perhaps more than usual in understanding geographical literature. Many of the names are rare or unique and may be presented in different ways in different texts (or even within the same text). Moreover, in the case of Avienus there is the additional problem that he was writing in Latin about a Greek discipline and thus had to transfer the names into Latin, perhaps not always accurately. Names may have come into Greek or Latin from another language, which is not always apparent. Especially in the text of Hanno, there was a tendency to substitute familiar Greek words for foreign names, which often gave them different meanings.[23] Geographical names would often be transmitted orally or through different languages before being written down, thus resulting in variant forms; this problem persisted into the nineteenth century and is rampant in western North America. Additionally, the same toponym will be different in Greek and Latin, with neither necessarily the "correct" form. Furthermore, Hanno's text is based on a lost Carthaginian original, and parts of Avienus' treatise may be as well. In this volume names are presented as closely as possible to their forms as provided by each author, although it is sometimes not possible to recover the nominative with certainty if it appears in no ancient text. As usual, some familiar names such as Carthage, Athens, and Rome are given in their English forms.

Notes

1 Homer, *Odyssey* 5.228–81; Lionel Casson, *Ships and Seamanship in the Ancient World* (Princeton 1971) 217–19.
2 Casson, *Ships* 30–2.
3 Herodotus 6.95.
4 *FGrHist* #709, T1.
5 Herodotus 4.44.
6 Hanno, *Periplous* 1.
7 Thucydides 1.24.1.
8 Pascal Arnaud, "Ancient Mariners Between Experience and Common Sense Geography," in *Features of Common Sense Geography: Implicit Knowledge Structures in Ancient Geographical Texts* (ed. Klaus Geus and Martin Thiering, Berlin 2014) 39–68.
9 For the history of *periploi*, see the accounts of various examples of the genre in J. Oliver Thomson, *History of Ancient Geography* (Cambridge 1948), and Duane W. Roller, *Ancient Geography: The Discovery of the World in Classical Greece and Rome* (London 2015), as well as M. Cary and E. H. Warmington, *The Ancient Explorers* (Baltimore 1963).
10 For the extant fragments of these travellers, see *FGrHist* #133 (Nearchos), #134 (Onesikritos), #711 (Androsthenes), and #712 (Patrokles).
11 Aristotle, *On the Heavens* 2.13.294a; Seneca, *Natural Questions* 3.14.1.
12 Diogenes Laertios 8.48, 9.21.
13 Eratosthenes, *Geography* F25; Thomson, *History* 110–12.
14 Posidonios F49 = Strabo, *Geography* 2.2.2.
15 Aristotle, *On the Heavens* 2.14.298b. Despite numerous attempts to create modern equivalents for the Greek stadion, the unit is so variable that such efforts are to a large extent futile, although very roughly one can say that there are about five stadia to a kilometer. See further, Aubrey Diller, "The Ancient Measurements of the Earth," *Isis* 40 (1949) 6–9; Duane W. Roller, *Eratosthenes' Geography* (Princeton 2010) 271–3.
16 Roller, *Eratosthenes' Geography* 263–7.
17 Aristotle, *Meteorologika* 2.5.362b.
18 This is the number of toponyms in the *Geographical Guide* of Ptolemy of Alexandria, published around the middle of the second century AD.
19 Aubrey Diller, *The Tradition of the Minor Greek Geographers* (New York 1952) 3–10; Hans Armin Gärtner, "Marcianus [1]," *BNP* 8 (2006) 304–5.

20 S. Gelenius, *Arriani et Hannonis Periplus* (Basel 1533); Jerker Blomqvist, *The Date and Origin of the Greek Version of* Hanno's Periplus (Lund 1979); Hanno, *Periplus* (ed. Al. N. Oikonomides and M. C. J. Miller, Chicago 1995); Duane W. Roller, *Through the Pillars of Herakles* (New York 2006) 129–32.

21 Diller, *Tradition* 102.

22 Didier Marcotte, *Les géographes grecques* 1 (Paris 2002) cxliv–clxiv; Diller, *Tradition* 165–76.

23 A good example from a later era of this toponymic slippage is the name Virginia, the North American colony and state, which was named after a local chieftain, Wingina, but was appropriated by Queen Elizabeth I as an honorific that also referred to the unspoiled nature of the land as well as her alleged personal characteristic, thus becoming far removed from the original eponym: see George R. Stewart, *Names on the Land: A Historical Account of Place-Naming in the United States* (New York 1945) 22.

1

CARTHAGINIAN EXPANSION AND EXPLORATION

The history of Carthaginian exploration—which mostly took place in the western Mediterranean and Atlantic regions—has been largely preserved through Greek and Latin sources, and thus in effect has become part of the Classical experience, in many ways inseparable from the accounts of Greek and Roman travellers to the same regions. In the last quarter of the ninth century BC (traditionally 814 BC), or somewhat earlier, Phoenician settlers, probably from Tyre, established a new outpost in North Africa at an easily defended site at the head of a deep bay about 60 km. southeast of the northernmost point of the continent.[1] Phoenicians had explored much of the Mediterranean, and were gaining a reputation as the primary seafarers of the region, "famous for their ships."[2] Their settlement was called Qart-Hadasht ("New City"), which became Karchedon in Greek and Karthago in Latin. It became the most important Phoenician city in the western Mediterranean, and by the sixth century BC, as the Phoenician homeland faded due to Assyrian and Babylonian pressure, Carthage began to take on an independent identity.

Carthage continued the Phoenician tradition of expansionism, so much so that it became necessary to define the limits of spheres of influence between it and the Etruscans.[3] Carthaginian settlements were established on the western Mediterranean islands and in the Iberian peninsula. But the great unknown was outside the Mediterranean, beyond the natural features that the Greeks called the Pillars of Herakles. There had been Phoenician probing into what lay beyond, into the great External Ocean, and it was said that they had established Gadir (Greek and Latin Gadeira or Gades, modern Cádiz), on the Iberian coast of the Atlantic, at about the time that Carthage was founded.[4]

Yet in Phoenician times there had been no systematic exploration of the Atlantic beyond its nearest portions. At first the Carthaginians did not venture into the outer ocean, concentrating their efforts in the western Mediterranean. They founded Panormos (modern Palermo) in Sicily, which had the finest harbor on the island, and other western Sicilian towns in the seventh century BC.[5] There had been Carthaginian settlement on Sardinia even earlier,[6] and the island came firmly under their control, as did Corsica

DOI: 10.4324/9781003030379-2

and the Balearics. There was also a presence in southwestern Iberia by 700 BC, but only south of the Tader River (modern Segura, located between Cartagena and Alicante).[7] With Carthaginian settlements in western Sicily and the other western Mediterranean islands by the late sixth century BC, as well as southwestern Iberia, Carthaginian expansionism in the Mediterranean seems to have come to an end. They considered the European mainland: from Iberia north of the Tader into the Keltic and Ligurian territory between the Pyrenees and Italy, and even Italy itself, but they did not move into these regions, in all likelihood because of the presence of two other major powers on the European coast: the Etruscans and various Greek states.

The Etruscans had expanded beyond their original homeland of Etruria (essentially modern Tuscany) as early as the beginning of the ninth century BC. They went south to the Tiber and north to the Padus (modern Po), and by the following century had established themselves around the Bay of Naples and had also made trading contacts throughout the western Mediterranean, gaining a foothold at least on Sardinia. They were a notable sea power and innovative in their shipbuilding.[8] It is probable the Carthaginians believed that the Etruscans had made it impossible for any settlement on the Italian peninsula.

There was also another presence in the western Mediterranean from at least the eighth century BC. Various Greek populations, beginning with those from the Central Greek island of Euboia, established settlements in southern Italy and Sicily. On the mainland these extended from Brentesion (modern Brindisi) and Taras (modern Taranto) in the east around to Kyme (modern Cuma) at the north end of the Bay of Naples. On Sicily, Greek towns were all along the east coast of the island, and west to the edge of the Carthaginian region. Selinous, at the southwest, was the westernmost Greek outpost, not far from the Carthaginian cities on the west coast. It was founded in the second half of the sixth century BC.[9] Moreover, it was only a short distance across the Sicilian Strait—the narrowest part of the Mediterranean—from Carthage itself. With this Greek presence in Sicily firmly established by 600 BC, there was nowhere on the island for the Carthaginians other than the narrow strip on the west that they already possessed.

There remained the Keltic-Ligurian coast and northern Iberia, between the Etruscans in northern Italy and the Tader River. But this too became closed to Carthaginian interests. Around 600 BC, emigrants from the Ionian city of Phokaia established Massalia (modern Marseille). They made a reconnaissance of much of the western Mediterranean, but Massalia was the center of their interests and soon became the most important Greek city west of Italy.[10] Other settlements by the Phokaians and Massalians soon populated much of the northwestern Mediterranean coast between the Etruscans and the Carthaginians.

Thus by the end of the sixth century BC, the Carthaginians found the western Mediterranean closed to new endeavors. Even their existing cities were threatened, by the Greeks in Sicily and the Etruscans in Sardinia. The only direction that remained available for trade opportunities and new settlement was to the west, through the Pillars of Herakles and outside the Mediterranean. Therefore plans were laid to send at least two major expeditions beyond the Pillars.

There had been some knowledge of what lay outside the Pillars for about a century before the Carthaginian expeditions. Gadir (modern Cádiz), lying just a few kilometers from the western outlet of the Mediterranean, had been founded by 800 BC, and there is evidence that the Phoenicians went farther along the western Iberian coast.[11] The Phoenicians may also have probed south of the Pillars.[12] But with the exception of Gadir these were merely reconnaissances, not the establishment of permanent settlements. A Greek expedition under the command of Kolaios, from Samos, had gone as far as the rich district of Tartessos in southwest Iberia about 630 BC, claiming to be off course but more probably using this as an excuse to investigate Phoenician or Carthaginian interests in the region.[13] This led to trade connections between the eastern Greeks and Tartessos, but Greeks seem to have had no further concern about what lay beyond the Pillars.

There was also the Phoenician circumnavigation of Africa during the reign of the Egyptian king Necho (610–595 BC). He sent an expedition around the continent in a clockwise direction, from the Red Sea to the Pillars of Herakles.[14] Opinion is still divided, and has been since antiquity, as to whether this journey actually took place, or was even completed. Herodotus' account provides details that support its existence—such as stopping to plant and harvest crops or changes in the celestial phenomena in the far south—but the journey, organized in Egypt, may have had little effect on Carthaginian aspirations a century later. Thus before around 500 BC there had been no extensive exploration of what lay beyond the Pillars or any broad conception of the western coasts of the African or European continents.

How much the Carthaginians knew about these early efforts is not certain. Nevertheless, around 500 BC the decision was made at the highest levels of the Carthaginian government to mount at least two major expeditions, one to the north of the Pillars and the other to the south. They would be led by two important members of the Carthaginian aristocracy: Hanno (Hannon) to the south and Himilco (Himilkon) to the north. Who these personalities were is not certain, since both names are among the most common known at Carthage.

The contemporary voyages may be seen as parallel events, since they served similar needs within Carthaginian policy. But the evidence for them is quite different. There is the extant summary of Hanno's cruise, derived from his own writings; the journey is also mentioned in a number of ancient sources, including the major ones on the topic of geography. On the other hand, Himilco is not mentioned in extant literature before the first century AD. A published report, now lost, still existed, which Pliny the Elder saw.[15] The major source for the expedition is material buried in the *Ora Maritima* of

Avienus, to be discussed later in this volume, and thus details of Himilco's expedition must be reconstructed from these sparse accounts and other equally elusive ones. He seems to have gone as far as Brittany, perhaps even to Ireland and the Azores. By contrast Hanno, as the extant text shows, went south into the West African tropics. These voyages, especially that of Hanno, were not without competition: at about the same time, the Massalians sent out their own Atlantic explorer, Euthymenes, who is also only documented in fragmentary and late sources, and like Himilco is not cited before the first century AD.[16] The use of the first person (*navigavi*) to describe Euthymenes' voyage demonstrates that, as in the case of Himilco, an actual report still existed which is now lost. Nevertheless it can be determined that Euthymenes went far south along the Atlantic coast of Africa and was probably the first Greek to report on crocodiles and hippopotami, thus reaching one of the major rivers of the tropics. But details of chronology are obscure enough that it cannot be determined whether Euthymenes precipitated Hanno's expedition or was in reaction to it.

Notes

1   Hédi Dridi, "Early Carthage: From its Foundation to the Battle of Himera (ca. 814–480 BCE)," in *The Oxford Handbook of the Phoenician and Punic Mediterranean* (ed. Carolina López-Ruiz and Brian R. Doak, Oxford 2019) 141–6.
2   Homer, *Odyssey* 15.415.
3   Aristotle, *Politics* 3.5.10.
4   Velleius Paterculus 1.2.3; Michael Dietler, "Colonial Encounters in Iberia and the Western Mediterranean: An Exploratory Framework," in *Colonial Encounters in Ancient Iberia* (ed. Michael Dietler and Carolina López-Ruiz, Chicago 2009) 7.
5   Thucydides 6.2.3; Diodoros 22.10.4; V. Tusa, "Panormos," *PECS* 671.
6   D. Manconi, "Sulcis," *PECS* 866–7.
7   Maria Carme Belarte, "Colonial Contacts and Protohistoric Indigenous Urbanism on the Mediterranean Coast of the Iberian Peninsula," in *Colonial Encounters in Ancient Iberia* (ed. Michael Dietler and Carolina López-Ruiz, Chicago 2009) 91–2.
8   Giovannangelo Camporeale, "The Etruscans and the Mediterranean," in *A Companion to the Etruscans* (ed. Sinclair Bell and Alexandra A. Carpino, Chichester 2015) 73–80.
9   Thucydides 6.4; Diodoros 13.59.4; V. Tusa, "Selinus," *PECS* 823–5.
10 Justin, *Epitome* 43.3; Athenaios 13.576a; John Boardman, *The Greeks Overseas: Their Early Colonies and Trade* (fourth edition, New York 1999) 216–19.
11 Ana Margarida Arruda, "Phoenician Colonization On the Atlantic Coast of the Iberian Peninsula," in *Colonial Encounters in Ancient Iberia* (ed. Michael Dietler and Carolina López-Ruiz, Chicago 2009) 113–30.
12 M. Euzennat, "Lixus," *PECS* 521; María Eugenia Aubet, *The Phoenicians and the West* (tr. Mary Turton, second edition, Cambridge 2001) 297–304.
13 Herodotus 4.152; Duane W. Roller, *Through the Pillars of Herakles: Greco-Roman Exploration of the Atlantic* (New York 2006) 3–6.
14 Herodotus 4.42; Roller, *Through the Pillars* 23–6.
15 Pliny, *Natural History* 1.5; 2.169.
16 Seneca, *Natural Questions* 4a.2.22; Roller, *Through the Pillars* 15–19.

2

THE *PERIPLOUS* OF HANNO

A brief text purports to be the *Periplous* of Hanno (Hannon), King of the Carthaginians. The extant Greek version is only a summary of a longer account, whether originally in Greek or Punic. As it stands, it is the earliest example of the genre preserved in Greek. Except for this text, the name of Hanno does not appear in Greek literature until Hellenistic times,[1] but there is abundant evidence that the Greek world knew about Carthaginian exploration beyond the Pillars of Herakles as early as the fifth century BC.[2] Thus it seems evident that Hanno's voyage was at that time or earlier, a conclusion also supported by linguistic analysis of the Greek version of the *periplous*.

The succinctness of the extant text does not hide its importance in the history of exploration, especially in terms of ancient knowledge of the West African coast. In fewer than 700 words it recounts an extraordinary voyage from Carthage through the Pillars and along the coast as far as the tropics, a distance of at least 7,000 km. The veracity of the document, and indeed the very existence of the journey, have long been questioned, but, if for no other reason, the sparseness and simplicity of the account give it credibility, since it is totally lacking in the fantastic phenomena typical of fictional narratives of exotic places.

The tradition of Hanno's *Periplous*

By the best evidence, the voyage of Hanno was before the second half of the fourth century BC, and probably substantially earlier. Although he was not mentioned by name until the Aristotelian *On Marvellous Things Heard*, of uncertain Hellenistic date, there is a steady stream of evidence that places the cruise around 500 BC. The extant Greek translation was composed in the fifth century BC.[3] Herodotus, writing no later than the 420s BC, knew about a major Carthaginian trade emporium on the West African coast, which may be one of the places founded by Hanno, demonstrating a Carthaginian presence in that region by that time.[4] A *periplous* by an author whose name is

DOI: 10.4324/9781003030379-3

Map 2.1 The voyage of Hanno.
Source: Map by E. Rodriguez

not known today but who is commonly referred to as Pseudo-Skylax, reports on travel down the west coast of Africa. He, or his source, visited Kerne, one of the places established by Hanno, and found it to be a flourishing trading center with commodities such as perfumed oil, precious stones, and Attic ceramics, all available at a major local festival, where tents were set up to promote the wares. This *periplous* is precisely dated to the 340s BC, although some of its material may be previous to that time.[5] Thus this work provides a solid *terminus ante quem* for Hanno's voyage, but it was certainly much earlier.

In Hellenistic times there was knowledge about the places that Hanno had visited, but often without any connection to the explorer himself. Eratosthenes' *Geography*, written in the second half of the third century BC, referred to Kerne but did not mention Hanno.[6] When Carthage fell to the Romans in 146 BC a number of Carthaginian documents were discovered and translated. Whether these included Hanno's full text is not known, but it may not be a coincidence that the historian and adventurer Polybios, present at the time, was given a commission by the Romans to sail down the west coast of Africa, replicating Hanno's voyage and visiting Kerne.[7] But again there is no mention of the Carthaginian in the extant material relating to this expedition.

The voyage of Hanno returned to general knowledge in the late first century BC due to the scholarly king of Mauretania, Juba II, who, with his wife Kleopatra Selene (the daughter of Kleopatra VII and Marcus Antonius), ruled northwest Africa at that time. Among Juba's several works—none of which is extant except in fragments—was one titled *The Wanderings of Hanno*. Since Juba ruled a part of the territory Hanno had explored, he probably had an interest in his efforts, and the work may have been an early composition, parts of which were later incorporated into Juba's major ethnography about Africa, *Libyka*. The title *Wanderings of Hanno* is only provided through a sparse reference in the *Deipnosophistai* of Athenaios, but vestiges of the treatise may appear in the *Natural History* of Pliny. Whether Juba had access to Hanno's complete text—rather than merely the extant Greek summary—cannot be determined.[8]

Nevertheless Juba seems to have played an important role in reawakening awareness of Hanno and his journey, and thus the major Roman geographers–Pomponius Mela and Pliny the Elder–were well aware of his voyage. Pomponius Mela provided a summary of his expedition, with Hanno one of the few geographical authors that he cited by name. Pliny found Hanno a major source for West African geography.[9] Exactly how Pomponius Mela and Pliny accessed Hanno's report, and in what form they saw it, remains unknown. The summary surviving today does not include all the information on the journey that was known in antiquity, and the material in later authors does not always agree with the extant text. Pomponius Mela reported that the countryside was deserted, seemingly a contradiction to the existing report, which may be proof that it was heavily condensed and that encounters with the locals were less frequent than they appear. Pliny recorded that the expedition started at Gadir and was designed to circumnavigate the continent of Africa, and referred to commentaries (*commentarii*) of Hanno, which suggests that a third rendering of the account was available in his day (in addition to the extant summary and Juba's analysis). Arrian noted that the cruise took 35 days; the extant text has 31 ½. These references show that other versions circulated as late as the Roman period, perhaps derived from Juba. If the full text was among the documents found at the fall of Carthage, it has vanished without trace. Yet the original would have been an example of the rich Carthaginian literary tradition, hardly known today.[10]

There has been a large amount of discussion in modern times as to whether the voyage actually took place, or whether the extant text is nothing more than a fabricated narrative, in the style of the fantasy geography genre that was popular in late Hellenistic and Roman times. Arguments against the voyage having occurred have generally been based on the winds and the lack of physical evidence in West Africa.[11] Yet a voyage that was largely a reconnaissance in its southern portions would leave no physical traces, and the establishment of trading centers such as Kerne, visited by others, is evidence in itself. That winds would have been adverse on the return cannot be denied, but this would hardly have been known at the beginning of the expedition, and tacking was a regular feature of ancient seamanship and is well documented.[12] Once Hanno and his companions reached their terminal point, somewhere in tropical West Africa, it is obvious that they would make every effort, however difficult, to return home, which they did successfully.

It can also be rejected that Hanno's report belongs within the genre of fantasy geography. Since the extant translation was created in the fifth century BC, it could not have been a Hellenistic or Roman fabrication. Certainly such a genre did flourish, beginning perhaps with Plato's Atlantis and other examples in the fourth century BC.[13] The best-known example, from a later period, is Plutarch's *Concerning the Face that Appears on the Globe of the Moon*. But these accounts have a moral tone and are a receptable for social criticism. Moreover, they are replete with strange encounters and anatomically impossible people. They can be based on real situations—Plutarch probably used Hanno as one of his sources (Carthage plays an important role in the narrative)—but this does not eliminate their fantastic nature. This element is totally lacking in Hanno's report: it is dry and straight-forward, almost to the point of being dull. The author has resisted any tendency to elaborate, and the text has the nature of a prosaic government document. Any suggestion that the voyage did not take place and that the account lacks credibility is totally implausible.

In summary, the extant text records that Hanno set forth from Carthage with a large fleet and allegedly 30,000 companions, and sailed through the Pillars of Herakles, heading in a southerly direction along the African coast. The expedition established a number of outposts (seven are mentioned), and at least one religious sanctuary. The first people encountered, the Lixitai, were friendly, but those beyond were increasingly hostile. Eventually they reached a point where they founded Kerne, their most important settlement. They passed a number of large rivers, and began to notice tropical flora and fauna. Several days beyond Kerne they turned back, but later continued on to the south; this may indicate a change in the purpose of the journey, from founding settlements to exploration. By this time the personnel of the expedition would have been seriously depleted, due to the foundations, and probably only a limited contingent went south of Kerne the second time. The local languages became unintelligible and the coast more mountainous. Eventually they came to the topographical feature known as the Horn of the West, and then entered a stretch of active vulcanism, although some of the fires that they observed were due to the locals. They passed the active volcano known as the Chariot of the Gods, and then, leaving the volcanic region, came to the Horn of the South. Here they encountered the hostile locals called the Gorillai, capturing three females and taking their skins back to Carthage. At this point the text seems to break off with the terse statement that supplies were running low and they could go no farther, a comment more enigmatic than informative.

The Greek text

The several ancient sources that refer to Hanno and his cruise need not have had access to the existing Greek text. As noted, varying versions of the report, including Juba's *Wanderings of Hanno*, circulated by late Hellenistic times. Nevertheless the translation was known by the mid-fourth century BC, when Ephorus used it.[14]

Although the *periplous* is a translation of a Punic original, it falls fully within the tradition of Greek literature.[15] As an early prose text, it has a number of poetic words and even vestiges of meter (not that the translator wrote in meter, but was accustomed to metrical expression).[16] This is indicative of a date when certain literary genres had only recently begun to use prose rather than poetry. The construction of sentences is remindful of some of the original Greek prose writers, such as Hekataios of Miletos, Pherekydes of Athens, and Akousilaos of Argos. As such, the Greek version of the *periplous* is probably one of the earliest examples of Greek prose. It is possible to conclude that it was created in the fifth century BC, perhaps earlier rather than later, and thus it has the first extant citation in prose of many Greek words.

Hanno and the Greek Language

In addition to the various words preserved for the first time in Greek prose, the *periplous* has a number of Greek personal and geographical names, such as Herakles, Kronos, Thymiaterion, Poseidon, Troglodytai, Theon Ochema, the Horns of the West and South, and, perhaps most notably, Aithiopes. It has long been assumed that these names were the work of the translator, who substituted Greek ones for indigenous or Punic ones. In some cases this is certainly true: the three personal names, Herakles, Kronos, and Poseidon, are presumably Greek renderings of Melqart, Baal Hammon, and the maritime Baal of Berytos. Thymiaterion, Troglodytai, the two Horns, and Theon Ochema are descriptive Greek terms (censer, Cave Dwellers, and Chariot of the Gods). The Horns (*keras*) are described by a Greek topographical term, and their directional positioning is also expressed in Greek. These were most likely due to the translator, who probably also converted Carthaginian distances to Greek stadia. But none of this is certain, and the name Aithiopes, or Aithiopians, is far more nuanced.

The Aithiopians appear twice in the *Periplous*: first in Section 7, where they are an inhospitable population living on the upper Lixos River. They were encountered again over 12 days later (Section 11), but these Aithiopians spoke a different language from those on the Lixos, in fact one that was incomprehensible to the Lixitai interpreters. It seems peculiar that there should be two groups of Aithiopians some distance apart speaking mutually incomprehensible languages. One can only presume that Hanno used the same ethnym for both even though they seem to be two distinct groups of people remote from one another.

Yet such a diverse nature of the Aithiopians, a population spread along the Atlantic coast of West Africa, conforms exactly to early Greek usage. Near the opening of the *Iliad* Zeus has gone to feast with the Aithiopians on the Ocean in the territory called the Land of the Aithiopians. They were so widespread that they were thought to be divided in two, extending from sunrise to sunset.[17] By the fourth century BC Ephorus believed that they occupied the entire southwestern quadrant of the inhabited world, the very region in which Hanno was said to have encountered them.[18] Granted Ephorus was over a century after Hanno, but the trajectory of thought seems clear: the Aithiopians were described essentially as Hanno did in his *periplous*, a population spread along the Atlantic coast of West Africa, divided into more than one part.

The obvious explanation—although purely speculative—is that Hanno knew the Homeric usage of the ethnym, and used it to name otherwise unknown populations scattered along the coast which seemed to fit the characterization of Homer. Thus Aithiopes was not a construct by the Greek translator for a Punic ethnym, but the very term (in a Punic form) that Hanno had used.

There is no reason that the Homeric poems could not have been known in Carthage in Hanno's day. Greek and Carthaginian territory had adjoined one another since the founding of Kyrene around 630 BC, and at the same time seamen such as Kolaios of Samos were probing Carthaginian regions. Around 540 BC, Carthaginians and Phokaians had fought off the coast of Corsica.[19] Thus by the time of Hanno Carthaginians were well aware of the Greek world, and may have wished to learn more about it. Any Greek attempting to acquaint Carthaginians with Greek culture would necessarily have begun with Homer. To be sure documented knowledge of the Greek language by Carthaginians is from a much later period. When Carthage fell to the Romans in

146 BC, the city was known to have had multiple libraries, whose collections the Romans were careful to preserve.[20] Librarians always seek to acquire as many books as possible, and it is a reasonable assumption that the librarians at Carthage obtained some Greek literature, having been aware of Greek culture since the seventh century BC.

Unfortunately there is no evidence for when the Carthaginian libraries were founded. But by the third century BC, and probably earlier, educated Carthaginians were fluent in Greek and were writing in that language. The most famous Carthaginian, Hannibal, had an entourage of Greek scholars who were with him for many years, including Sosylos of Lakedaimon, who taught him Greek literature, and also Silenos of Kale Akte. Moreover, Hannibal wrote a number of treatises in Greek; although the topics of most of these are unknown, they included an address to the Rhodians warning them to beware of the Romans, written just before his death in 183 BC.[21]

There is no indication when Carthaginians began to read and write in Greek, but this raises the possibility that Hanno himself knew the language and actually used Homeric imagery in the report of his cruise. Moreover, Hanno or his associates may have been the ones to commission the Greek summary of the report, realizing that Greek knowledge of the great achievements of Carthage would be of advantage to both states.

The translation of the *Periplous* of Hanno

The following translation is based on the edition by Al. N. Oikonomides and M. C. J. Miller of 1995 with some minor adjustments by the present translator.[22] The enumeration of sections does not appear in the manuscript, but were the addition of C. Müller in the mid-nineteenth century.[23] There are three places where there seem to be gaps in the text, which are not noted in the manuscript but are an obvious interruption to the flow of the narrative. In addition, the text is disjointed with rough transitions (especially around Sections 10–12), and it is possible that the extant translation is, in part, a composite of more than one Carthaginian report.[24]

The text

The Periplous of Hanno, king of the Karchedonians, into the part of the Libyan land beyond the Pillars of Herakles, which he set up in the precinct of Kronos, and which revealed the following:

(1) It was decreed by the Karchedonians that Hanno sail outside the Pillars of Herakles and found Libyphoenician cities. Thus he sailed in command of 60 penteconters and with a large number of men and women, as many as 30,000 in number, and with grain and other provisions… .

18

(2) Having put to sea, we passed the Pillars and sailed beyond them for two days, and we established our first city which we named Thymiaterion. Below it there was a large plain.

(3) Then we put to sea, toward the west, and we came together at Soloeis, a Libyan promontory that was overgrown with trees.

(4) We founded a sanctuary of Poseidon there, and went back on board, sailing toward the sunrise for half a day until we came to a lagoon that was located not far from the sea. It was filled with many large reeds. There were elephants in it, with numerous other animals living there.

(5) Going past the lagoon, and sailing for about a day, we established cities near the sea, called Karichon Teichos, Gytte, Akra, Melitta, and Arambys.

(6) From there we put to sea and came to a large river, the Lixos, which flows from Libya. Along it were a nomadic people, the Lixitai, who grazed cattle. We remained among them for a while, becoming friends.

(7) Inland from there live inhospitable Aithiopians, whose grazing land was infested with wild animals and divided up by high mountains, from which, they say, the Lixos flows. Around the mountains live strangely shaped people, the Troglodytai, whom the Lixitai said were faster on a racecourse than horses.

(8) Taking interpreters from them, we sailed along a wilderness toward the south for two days...from there back toward the sunrise for the run of a day. We found a small island there in the recess of a certain bay, which was five stadia around. We settled there and named it Kerne. We estimated from our *periplous* that it lay on a line with Karchedon, for the cruise seemed to be the same from Karchedon to the Pillars as from there to Kerne.

(9) From there we came to a lagoon, and sailed on a certain large river, the Chretes. The lagoon had three islands in it, larger than Kerne. From there we sailed for a day into the recess of the lagoon. There were exceedingly large mountains extending above it, full of wild people dressed in animal skins. They threw rocks at us in order to sweep us from the ships and to prevent us from disembarking.

(10) Sailing from there we came to another large and broad river which was filled with crocodiles and hippopotami. We turned back from there and returned to Kerne.

(11) From there we sailed south for 12 days, keeping close to the land, all of which was inhabited by Aithiopians who fled from us and

did not remain. What they said could not be understood by the Lixitai with us.

(12) On the last day we anchored at high and wooded mountains. The wood of the trees was fragrant and multi-colored.

(13) Sailing around them for two days, we came to an deep gulf of the ocean with a plain toward the land on either side. When we looked at night there was fire rising up everywhere in the distance, some greater and some smaller.

(14) We watered there and sailed ahead for five days, always along the land, and came to a great bay which the interpreters said was called the Horn of the West. There was a large island in it, and on the island there was a lagoon of sea water. There was another island in it, on which we landed, seeing nothing in the daytime but woods. But at night many fires were burning, and we heard the sound of flutes and cymbals, the beating of drums, and an immense amount of shouting. We were taken with fear, and the seers ordered us to leave the island.

(15) We sailed away quickly and passed by a region that was full of fiery incense, from which streams of fire flowed down into the ocean. The land was inaccessible because of the heat.

(16) We quickly sailed away from there, because we were afraid, and continued for four days. We saw that at night the land was full of fire, and in the middle there was a fire higher than the others that seemed to touch the stars. In the daytime we could see that this was an exceedingly high mountain, which was called the Chariot of the Gods.

(17) Three days from there, sailing past streams of fire, we came to the bay called the Horn of the South.

(18) There was an island in its recess, like the first one, having a lagoon in it and another island within it, filled with wild people. Most of them were women, with hairy bodies, whom the interpreters called the Gorillai. Chasing them, we were unable to capture the men, since they all escaped by climbing the cliffs and defending themselves with rocks, but we took three women, who bit and scratched those taking them and did not want to follow. But we killed them and skinned them, and brought their hides to Karchedon... . Yet we did not sail any farther, since we were out of provisions.

Commentary

The titular introduction to Hanno's *Periplous* is placed separately before Section 1. It consists of four lines that identify the author, the purpose of the expedition, and the eventual disposition of the report, although in what medium is not specified. It is probable that this was an introduction added by the Greek translator in order to position the reader in regard to the document as a whole. The nomenclature has been hellenized. "King" (*basileus*) is not a term that the Carthaginians would have used to describe their leaders in this period: the proper word was *shopet* (suffete). Yet *basileus* would have been more meaningful to a Greek reader, and there is some evidence that there had been a king (*melek*) at Carthage in its earliest days, but the office had vanished by the time of Hanno.[25] Nevertheless Hanno held a high position in the Carthaginian hierarchy, perhaps even chief magistrate, although one might question the wisdom of sending such a person on a lengthy expedition to unknown regions. Given the frequency of the name, it is not possible to connect Hanno the explorer with any other known personalities.[26]

Karchedon (here the ethnym rather than the toponym) was the standard Greek form of the name of the city, which the locals called Qart-Hadasht. This, and other uses of the name in the *Periplous*, may be their earliest citation in Greek: both the toponym and ethnym were used frequently by Herodotus, and the toponym was known to Sophokles.[27]

This may also be the first use of *periplous* to mean a sea voyage: only the title of Skylax of Karyanda's work, from the late sixth century BC, may precede this, but its authenticity is dubious. Yet the word in this meaning is definitely documented by the 330s BC, and probably in Hanno's day was undergoing the change from merely "sailing around" to a definition of a literary genre.

Throughout the text, the translator generally used Greek geographical terms in place of any Punic originals. As was common in Greek sources, the toponym "Libya" meant the entire continent of Africa, although its extent was hardly known in classical antiquity. The term was probably originally limited to regions immediately west of Egypt and the citations in the *Periplous* may be the earliest Greek use in a continental sense.[28] Since the toponym was North African in origin, it may approximate the Punic term of the original text.

The Pillars of Herakles were, to the Carthaginians, the Pillars of Melqart, who was the local divinity of Tyre and whose cults came west with the Phoenician settlers. By the fifth century BC there was a certain amount of assimilation with the Greek Herakles, many of whose activities were placed in the western Mediterranean. When Herodotus visited Tyre he saw two pillars that were dedicated to Melqart (whom he called Herakles).[29] In time, they were manifested geographically at the other end of the Mediterranean. Greeks originally called the entrance to the sea the Pillars of Atlas, but the name evolved to the Pillars of Herakles, presumably with the partial assimilation of Melqart and the Greek hero. In Greek literature the change had taken place by the fifth century BC, since Herodotus only used the latter term. The text of Hanno's *Periplous* may be one of the earliest citations of the toponym.[30]

Kronos was the Greek name for the Carthaginian deity Baal Hammon, who, along with his consort Tinnet, played a prominent role in the divine hierarchy of Carthage. He was also of Tyrian origin, and remained important (as Saturn) into the Roman period.[31] His sanctuary would have been one of the major locales in the city. The text implies that there may have been a public inscription, although it is equally possible that the temple contained a depository of official records. The extant text is rather long for a public inscription, yet at the same time somewhat terse for an official report.

Section 1

The first section is the only part of the text in the third person, and establishes the legal basis of the expedition. There were two assemblies at Carthage, a senior one and a more popular one, and presumably one or both of these provided Hanno's commission, perhaps the latter, since the expedition seems to have had the general approval of the populace.[32] An important factor is that Hanno was instructed to go outside the Pillars of Herakles/Melqart; in other words, beyond the traditional boundaries of Carthaginian territory. The specific statement was probably to make his orders perfectly clear: it was a journey of 1,500 km. even to reach the Pillars from Carthage. They also explicitly stated that the purpose of the expedition was to establish new cities (the Greek word *polis* is used, but it is doubtful that the foundations were cities in any Hellenic sense). Yet the expedition somehow evolved, since no new foundations are mentioned after Section 8, and it became a voyage of exploration into regions that had probably not been previously seen by any Carthaginians. Like many explorers from ancient to modern times, Hanno interpreted his instructions as broadly as possible in order to reach new territory. The ethnym "Libyphoenician" may be a hellenized form of whatever the Carthaginians used to describe their ethnic nature, seeing themselves as Phoenicians who had come to Libya. There is no further extant citation of the term until the second century BC, so it does not seem to have entered the mainstream of classical diction until a much later date.[33] This may be proof that Hanno's translator created an ethnic neologism from a Punic original.

The text describes a massive expedition, with numbers that are improbable. There is no reason to doubt that Hanno used penteconters: they were serviceable ships that had their origins in the Bronze Age and were common throughout the ancient Mediterranean. They were the mainstay of Achilles' fleet when he went to Troy. The early references are descriptive; the technical term "penteconter" did not appear until the fifth century BC, with the citation in the *Periplous* one of the first uses of the word.[34]

The expedition was probably large—it founded at least seven outposts—but the numbers recorded seem excessive. A more reasonable size might be 5,000–6,000.[35] It was common to exaggerate the strength of forces—the number of those attacking Troy and Xerxes' million-man army come to mind—and although

Hanno's expedition was not military, there was probably the same tendency. There is no way to determine a reasonable amount of participants; other reports of early expeditions are wisely vague about numbers, such as the Indos cruise of Skylax of Karyanda or the Phoenician circumnavigation of Africa, where the reports refer only to "ships."[36]

The expedition, as expected, was well provisioned, with *sita* (singular *sitos*) specifically mentioned. The word can simply mean "food," especially that which is grain oriented, such as in the banquet Alkinoos had prepared for Odysseus, but it more generally means "grain"; an example is the statement made by Athena about the agricultural bounty of Phaiakia.[37] It is probable that Hanno took seeds with him in order to stock the new outposts; the use of the plural supports this. There was a precedent with the Phoenician circumnavigation a century previously, whose participants planned to plant while on route. Hanno was probably aware of this method of providing supplies for a lengthy expedition.

Section 2

Section 2 does not seem to connect directly with the previous one, and a gap in the extant text is possible. Most notable is the shift from third to first person, which remains for the rest of the account. In addition, the phrase "having put out to sea" (*anachthentes*) implies that the expedition is not at its beginning, but has stopped at a port after its departure from Carthage. This is the word used to describe the Persian expedition to south Italy returning to sea after a stopover at Kroton, or after Kolaios of Samos stopped at Platea.[38] It seems that the fleet put into port somewhere, mostly likely at Gadir. Pliny reported that this was its starting point, and it would be natural for the expedition, having sailed 1,500 km. from Carthage, to provision and water at the last major city in the Carthaginian world before striking out into little-known territory.[39] Such an assumption is strengthened by the phrase "we passed the Pillars" (*paremeipsamen*), rather than going between them, a good description of the route from Gadir to the nearest point in Africa, which would pass by the outlet of the Mediterranean. The sail from Carthage to Gadir might have taken somewhat over a week: in the first century AD the fastest sailing time from Gadir to Ostia, not much longer than the Carthage–Gadir run but in a world of faster ships, was in fact a week.[40] If Hanno's report included an account of that part of the expedition, over a well-travelled route, it might have been brief and routine, and ignored by the translator.

Assuming, then, that the fleet put into Gadir and resupplied, it then headed southeast along the Iberian coast, passed by the Pillars at the outlet of the Mediterranean, and struck the African coast around modern Tangier. The two days' sail is difficult to understand since it is not possible exactly to define what Hanno meant by the Pillars. Their location was astonishingly variable in antiquity, and they were placed throughout the strait between the Atlantic and the Mediterranean.[41] If Hanno were being specific, the most likely place for the Pillars in this context is the Hera (or Juno) Promontory, modern Cape Trafalgar, where the coast, having run southeast from Gadir for 45 km., turns to the east, beginning the straits into the Mediterranean. At this point Hanno would have seen the African coast 50 km. dead ahead, and, keeping on the same heading, would have reached it around the site of Tangier. The "two days" may be the sailing time from Trafalgar, which would be a slow speed, but the currents were adverse to the direction of travel.

Thymiaterion, the first city founded by the expedition, is probably Tingis (modern Tangier). It had a Greek mythical foundation story, and there are archaeological remains from perhaps the eighth century BC.[42] But there may have been little more than a Phoenician outpost at the site, guarding the south side of the strait. Hanno may have wanted to assert a Carthaginian claim over an earlier settlement and turn it into a proper town. The fact that Thymiaterion lay on a height overlooking a large plain conforms to the location of Tingis, where early remains are on a plateau northeast of the later port, and there is an extensive plain to the southwest. A *thymiaterion* is a censer, a vessel that could be dedicated at a shrine,[43] and the translator probably discarded any indigenous or Punic name (essentially Tingi). Hanno would have made a religious offering here, on the occasion of landing in northwest Africa, and the translator turned the local name into a Greek religious term.

Section 3

From Thymiaterion the expedition sailed west, following the coastline, which runs slightly north of west for about 10 km. from Tangier, and then turns sharply south. The suggestion that the fleet "came together" (*synelthomen*) at Soloeis indicates that it had become somewhat scattered, inevitable for a large fleet. Soloeis was a promontory that was heavily wooded, and after passing it the expedition made a sharp turn to the left (Section 4). This means that it has reached the northwest corner of Africa, modern Cape Spartel, which is still heavily wooded today. Soloeis is one of the few toponyms in the *Periplous* that can easily be identified, since it came to be recognized as the farthest extremity (from a Mediterranean point of view) of the continent of Africa.[44]

Section 4

At Soloeis the expedition established an altar to Poseidon, as the translator called it. This would be Baal in a maritime role, who seems to have originated in the Phoenician city of Berytos. A traveller saw the altar sometime later, by the 330s BC, and reported on its magnificence, with human and animal reliefs that were so impressive they were attributed to the mythical artist and craftsman Daidalos.[45]

The first major topographical problem in the *Periplous* follows. From Soloeis the expedition sailed "toward the sunrise" for a brief period. This would be a direction between northeast and southeast (depending on the season), and there is no place along the Atlantic coast of Africa that one can do this until rounding the Atlas mountains in the vicinity of modern Agadir, hundreds of kilometers south of Soloeis and even well beyond Lixos (Section 6). This has led to complex topographic convolutions by many modern commentators, often resulting in a virtual rewriting of the *Periplous*.[46] This seems unnecessary since there are simpler explanations. It is possible that a small portion of the text is misplaced, perhaps including the mention of elephants, but it could simply be an error, either in the original or translation. It may also be that the expedition had ended up rather far out to sea and had to head easterly to return to the coast.

The lagoon cannot be located beyond being somewhere in the 100 km. between Soloeis and Lixos. The word, *limne*, in this context means a marsh at the edge of the sea (or a river). Reeds (*kalamoi*) are common on the West African coast, and may have been sugar cane (*saccharum officinarum*); those from northwest Africa remained a curiosity into the Roman period, when ones are documented whose joints had a capacity of about eight liters.[47]

This is one of the earliest citations of the elephant in Greek. The species would be the now-extinct North African variety, discussed in detail by Juba II. Half a day south of Soloeis is far north for the elephant—around the western end of the Atlas mountains would be more reasonable—which reinforces the possibility that a portion of Section 4 is out of place.[48]

Section 5

This section is heavily condensed, and may be a summary of the outposts that were founded between Soloeis and Lixos. It is improbable that five places were established in a single day. Assuming that much of Section 4 is out of place (as well as the reference to the "lagoon" at the beginning of Section 5), these locations would be scattered at intervals along the coast southwest of Soloeis. Rather than use the traditional "we founded" (*ektisamen*), as for Thymiaterion, these were merely "established" (*katoikesamen*), a term referring more to a settlement or population adjustment than a formal city foundation. Among the examples are when Media was divided into settlements and the settling of Ionians and Karians near Memphis.[49] Hanno's expedition would have created small outposts—trading centers or forts—designed to examine the possibilities of trade in a given region. These would have been the first of many Carthaginian settlements that became scattered along this coast. At the peak of Carthaginian power—essentially the fifth through third centuries BC—it was said that there were 300 outposts in this region, probably an exaggerated number, but perhaps including many small forts and trading posts. By the late third century BC, as the local Carthaginian presence faded, it was reported that many were abandoned without trace, which also suggests an ephemeral nature.[50] Nevertheless it is plausible that Hanno dropped off detachments at regular intervals along the coast, in order to create small settlements that would reconnoiter the region.

Five are mentioned by name, and none can be specifically located. Three seem to have Greek names: Karichon Teichos, Akra, and Melitta. These are probably translations of Carthaginian toponyms, or the conversion into Greek forms of orthographically similar Carthaginian words.[51] Karichon Teichos may sound like "Karian Wall," but this is highly unlikely, although the use of "wall" may indicate a Carthaginian fortified settlement, with "Karichon" a hellenization of whatever word Hanno used. The toponym was cited by Ephorus, writing by the 330s BC, perhaps the earliest documentation of an awareness of the Greek translation of the *Periplous*.[52] Akra is a common Greek toponym applied to any site on a headland or height, and is probably a direct translation of the Carthaginian term. Melitta may be an original Punic toponym that sounded Greek ("honey"), or a translation; it was known to Hekataios of Miletos.[53] Gytte and Arambys seem legitimate local toponyms. The former may be the place known as Cotte to Pliny, which has been dubiously located near modern Jibila, 15 km. from Tangier and where there are fish-salting works from the period of Juba II. The latter has echoes of the Homeric Eremboi, or Aramboi, a Libyan population encountered by Menelaos, and Hanno or the translator may have adjusted the Carthaginian name appropriately.[54]

Section 6

Next the expedition came to the mouth of a major river, the Lixos, presumably the local name. This region became its headquarters for an extended period, and there were reconnaissances into the interior (Section 7). Lixos is the only site immediately south of Soloeis that can be identified without serious question, although, inevitably, there are problems of interpretation, largely due to the apparent erroneous placement of material in Section 4, which would give the impression that Lixos was farther south, into the tropics. The known site of Lixos is only 70 km. south of the northwest corner of Africa, not in the elephant country; yet it seems certain this is the place that Hanno visited.

The Lixos River is the modern Loukos, which preserves the ancient name. It is the first significant watercourse the expedition would have encountered after the Pillars of Herakles, and flows from the modern Rif mountains of interior Morocco. Hanno described it as coming "from Libya," suggesting that he believed he was no longer within the traditional confines of the continent. The river makes a broad arc to the south before entering the ocean at modern Larache, having passed through a fertile drainage. It is short (only 68 km.) but high in volume. The major archaeological site of Lixos lies on a bluff northeast of Larache and on the right bank of the river about 4 km. from its mouth.

It is possible that Hanno knew about this fertile region before leaving Carthage: the site was settled as early as the beginning of the sixth century BC, but there does not seem to have been any permanent occupation at that time. It became the best-known locality on this coast: the Greek world learned about it not long after Hanno's time, or even before. Its fertility led to the belief that it may have been the mythological Garden of the Hesperides, but this is not documented until late Hellenistic times and may have been promoted by Juba II as part of his royal ideology.[55] The twisting of the lower Lixos River, still conspicuous today, was said to represent the serpent who guarded the Apples of the Hesperides.

The visible remains at Lixos today are no earlier than the fourth century BC, and the town gained renewed prominence under Juba.[56] But when Hanno visited there was little, if anything, to see in terms of a permanent settlement; the emphasis is on the nomadic locals. In fact, the *Periplous* only mentions the river, no townsite. Yet in later times the toponym Emporikos was applied to this region, showing the vestiges of a Carthaginian mercantile presence that was probably founded, or at least systematized, by Hanno.[57]

The local Lixitai were hospitable, and the expedition remained for a while, presumably learning about the region and making a number of explorations into the interior (Section 7). They were described as nomads who grazed animals; the translator used the largely poetic word, *boskemata*, to describe their herds.[58] Their ability at running may be a misplaced hint of the characteristics of certain sub-Saharan peoples. They came to be remembered as one of the most remote western populations,[59] and there was a distinct contrast between their friendliness and the hostility of the peoples that the expedition would encounter thereafter, either inland or farther down the coast.

Section 7

During its stay at Lixos, the expedition may have sent scouting parties up the river, although some of the information was received orally from the local pastoralists (as indicated by "they say"). The mountains east and southeast of the coast rise to 1,595 m. in the Lalla Outka region near the source of the river; these are presumably the "high mountains" mentioned in the report.

At this point the expedition had the first of two encounters with the Aithiopians (see also Section 11). As noted previously,[60] this ethnym may have been Hanno's original term, if he knew the Homeric poems, or it may be that the translator—certainly aware of the epics—substituted Homeric ethnography for whatever name Hanno used. Regardless, the presence of the Aithiopians at two places in the *Periplous* remains an oddity in the account, and it is perhaps relevant that they remained elusive.

Like the Aithiopians, the Troglodytai are an ethnic peculiarity among the peoples of the *Periplous*. The ethnym is the standard Greek term for "Cave Dwellers," who lived in many places in the world. But there was also an

African ethnic group, the Trogodytai, who lived between the Nile and the Red Sea, and the two ethnyms have been confused since antiquity. There were also Trogodytai or Troglodytai who lived in the central Sahara.[61] "Troglodytai" is a broader and more generic term than "Trogodytai," and the manuscript of the *Periplous* distinctly has the former word. If the original Punic text used one of these words, it represents further knowledge by Hanno of Greek views of African ethnology. Yet it is also possible that the translator simply gave the Greek word for Punic "cave dwellers." Whether Troglodytai or Trogodytai, their unusual anatomical characteristics are a common Greek formula for remote peoples, but the Carthaginians may have believed the same. Yet of all the sections of the *Periplous*, this seems most connected to Greek ethnographic thought.

With Section 7, the *Periplous* becomes darker in tone, beginning with the Aithiopians, who were "inhospitable" (*axenos*), a largely poetic word, diction which perhaps—in the translator's mind—better expressed the remote world that the expedition was now entering.[62] The remainder of the text is largely a series of encounters with unfriendly people. Although it has been argued that from this point the expedition enters the world of fantasy geography, elaborated by the Greek translator from Hanno's data,[63] it is more likely that the explorers were moving into an alien world, beyond their limits of knowledge, where people were strange, spoke no known language, and had unusual practices. Similar feelings were expressed by the Persian explorer Sataspes in the same region of coastal Africa during the period 479–465 BC,[64] and, much later, the Roman propraetor C. Suetonius Paulinus, who crossed western Africa from the Mauretanian coast of the Mediterranean and perhaps reached the Niger River.[65] Hanno's reaction to the new world around him, as well as that of other explorers, was no different from most of those who ventured far beyond the limits of civilization from antiquity into recent times.

Section 8

At the Lixos River, the expedition left the known world and entered a region devoid (*ereme*) of people. Although this word is often translated as "desert" this is erroneous; the coast south of the river continues to be quite fertile as far as the Atlas Mountains. The proper meaning here is "deserted" or "desolate," as when Aigisthos banished the singer who had been left to guard Klytaimnestra to a deserted island, or, more relevant to Hanno's diction, the deserted country north of the Black Sea.[66] Hanno well realized that the cruise was moving into unknown territory, and took on board Lixitai interpreters (*hermeneiai*). This is perhaps the earliest citation of a Greek word that would become increasingly common in the expanding world of the fifth century BC: Herodotus needed interpreters in Egypt.[67] The Lixitai would have known Punic, which suggests that they had had previous contact with the Carthaginians.[68]

The expedition continued along the coast, but there is a gap in the text at this point, since no place appears to identify the meaning of "from there." From the Lixos River to Kerne was probably more than the three days preserved, and Kerne, the next toponym, was one day after the coast turned to the east ("sunrise"), remindful of the possibly misplaced Section 4. The textual gap makes speculation about this part of the *Periplous* almost impossible, although the first place that the coast turns to the east is at modern Agadir, where it goes around the west end of the Atlas range.

The name Kerne may mean "last settlement" or "horn."[69] If the former, it would imply the expedition planned to turn back from there, but that some of its members decided to go farther, although without making any settlements. Its location cannot be determined, yet every possible argument has been advanced to place it along hundreds of kilometers of coast. The *Periplous* of Pseudo-Skylax put it 12 days from the Pillars of Herakles or seven from Soloeis, but this is not too meaningful because the numbers in the text are uncertain and nothing is known about the expedition's sailing speed or time spent in port.[70] Most available data about ancient speeds are from Hellenistic and Roman times, when ships were faster and routes better known. The maximum speed recorded from that period is 6.2 knots,[71] but Hanno's fleet would hardly have gone that fast, and it also spent extensive time in port at the Lixos River. The variables are immense but an estimate of 1,100 km. for the 12 days is a possibility.[72]

The two most likely locations for Kerne are the estuary of the Río de Oro at modern Dahkla in the Western Sahara, and the Bay of Arguin, about 145 km. farther along the coast, just across the border from Morocco into Mauritania. Near Dahkla is the toponym "Herné" which is suggestive but perhaps not very meaningful, and may not be a survival from antiquity. The Bay of Arguin and Arguin Island within it seem a better fit, and are about 1,100 km. from the Pillars; the island corresponds somewhat to the description in the *Periplous*, and in early modern times it was a colonial trading post. The major argument against it is that it has no good anchorage today and is larger than the five stadia across (about 1 km.) reported, but these probably represent changes since antiquity. Yet all possibilities for the location of Kerne are speculative, and various commentators have placed it all the way from Mogador (Essaouria), north of Agadir and where Juba II established an important dyeing industry,[73] to as far south as the Senegal River, which seems improbable.[74] Such attempts to locate Kerne are more interesting than profitable and probably futile.

Nevertheless the trading post at Kerne became the best-known location of Hanno's cruise, and was visited occasionally by Greeks over the next several hundred years. It faded when Carthage declined, and had been long abandoned by the first century BC, although two centuries later Ptolemy in his *Geographical Guide* still knew the name, but hardly understood its location other than somewhere off the west coast of Africa.[75] Yet as long as it flourished it may have been the most important Carthaginian outpost on this coast. Herodotus preserved an interesting account of Carthaginian trade methods, which may apply to Kerne and certainly to other posts on the West African coast.[76] It was a silent process, with cargo unloaded and a fire lit, and then the traders returning to their ships. The locals would come and examine the goods, and leave gold in payment, which the Carthaginians would only take if they thought the amount was sufficient. If the gold was ignored the locals would add more, and the process might be repeated several times, but eventually the gold and the wares would be removed, with neither party having had contact with the other, something that avoided confusion if there were no common language.

The expedition spent an undetermined length of time at Kerne, creating a permanent settlement, which would mean that a number of Carthaginians were left there. Exploratory journeys were sent forth (Section 10). Kerne is the only place mentioned in the *Periplous* where there was an attempt to locate it relative to Carthage: since it was such an important foundation, there may have been a certain anxiousness for the travellers to have some sense of where they were.[77] The methodology used was rough, based primarily on data from the ships' logs, not any precise measurement. Currents and winds would have affected sailing speed, and Hanno and his companions had little way of determining these. Yet they wanted some sense of where they were, especially if it were thought (for a while, at least), that this would be the terminal point of the expedition. Suggesting that Kerne was on a line with Carthage is less informative than it sounds, and indeed is somewhat ambiguous (since any two points would be on a line), but the statement is explained by the rest of the sentence, that Kerne was as far from the Pillars as they were from Carthage. It was more a psychological comment for the benefit of the expedition members than a geographical one.

Section 9

Presumably some if not most of the party remained at Kerne and became engaged in the process of establishing the outpost, while a small detachment, perhaps two or three ships, was sent farther down the coast to the south. Most of the remaining part of the expedition (26 of the 31 ½ days of the total cruise) has few details or toponyms, a characteristic of exploration in remote unknown regions. Yet the vagueness indicates the validity of the report, however confused the extant text may be.

Hanno and his companions reached a large river and the territory of the hippopotamos (Section 10). Wherever Kerne was, this means that the group had passed the deserted and inhospitable coast of modern Mauritania and entered the *sahel*, the transitional zone between the desert and the tropics. They came to the first of the large tropical rivers of West Africa, the Senegal, which was called (according to the manuscript) the Chretes. This is probably the river known to Aristotle as the Chremetes, which he believed was one of the largest in Africa.[78] If the Chretes is the modern Senegal, it provides a better fixed point for the expedition than Kerne. The river is 1,086 km. long and has its origins in the highlands of Guinea before flowing through Mali and into Senegal. The major problem in identifying it with Hanno's Chretes or Chremetes is that there are no high mountains a day's sail upriver. But there seems no other choice, and its outlet (at modern St.-Louis) is marked by a lagoon and islands. It is possible that there is another gap in the text and the mountainous wild people are near another river, yet this entire coastal region (from modern Senegal to western Guinea) is one of large rivers and estuaries, not mountains.

The expedition sailed up the river for an unknown distance, and encountered unidentified wild stone throwers who dressed in animal skins. Although the original Homeric meaning of *aperaxan* is "to crush," by the fifth century BC it came to have a specific nautical context, "to sweep from the decks of a ship," which happened at Salamis.[79] This is probably what occurred on the Chretes River; it is unlikely that the locals handled stones large enough to crush people. Presumably the Carthaginians turned back downriver at this point.

Section 10

After the Chretes, the expedition continued to another large river, which is not named. If the Chretes is the Senegal, this would be the Gambia, about 400 km. to the south, which originates in the highlands of Guinea and flows 1,120 km. to the Atlantic, emptying into the ocean in a broad estuary. In the unnamed river the expedition encountered crocodiles and hippopotami. Crocodiles were already known to the Greek world from the Nile, yet what Hanno saw was a different species (probably *Ostelaemus tetraspis* rather than *Crocodylus niloticus*).[80] This is probably the first citation in Greek of any type of the animal.

The hippopotamos is also first mentioned here, although it too existed in the Nile. The translator would have taken a Punic or indigenous term and rendered it into the Greek "river horse." Which of the two species of the animal—the common variety, *Hippopotamus amphibius*, or the pygmy hippopotamus, *Hexaprotodon liberiensis*—was encountered is not known; geography favors the latter.[81]

The *Periplous* is exceedingly laconic at this point, with no ethnic details or toponyms, and includes the brief statement without explanation that the expedition returned to Kerne. Establishing any kind of settlement no longer seems to have been a goal, and this supports the idea that the sail to the two rivers was merely a reconnaissance from Kerne. The itinerary becomes increasingly confused from here into Section 13, with contradictory data about sailing times and the purpose of the expedition. It is possible that some of this material conflates more than one cruise.[82]

One reason for the sailing back and forth in these regions may have been an attempt to learn if there were others from the Mediterranean in the vicinity. The expedition of Euthymenes of Massalia visited the region of the Senegal River at about the same time as that of Hanno—although it is impossible to determine which was first—and the Carthaginians, if they heard rumors about Massalian activity in the area, may have wished to make a prior claim.[83] The rather sudden ending of Section 10, with the unexplained return to Kerne, suggests that the expedition itself was at an end, but if the material in Sections 11–18 is not from another cruise, a decision must have been made upon return to send out a party that would continue farther than the points already reached.

Section 11

A sail of 12 days, reported without providing any details, is the largest travel span of the *Periplous*, over ⅓ of the total amount of the account. It seems implausible that the expedition found virtually nothing worth mentioning, and this may be the summarization of the translator. Contact with the locals was minimal. They were still generically described as "Aithiopians," but the Lixitai interpreters could not understand them. This means that the expedition has passed a language frontier, which was at the Senegal River, and that it has moved well to the south.[84] Like Section 10, Section 11 is terse and brief, with no topographical data.

Section 12

This is the most enigmatic section of the *Periplous*. Either it is badly misplaced (perhaps belonging at the end of Section 18, before the final sentence), or it represents the end of one portion of the expedition. Section 13, with its additional two days of sailing, directly contradicts the "last day" of Section 12, but the two join together with the mountains of the former referenced in the latter. It is probable that the expedition is somewhere along the coast from Guinea to Liberia, where the mountains come closer to the sea than at any location since the Atlas. The fragrance and floral color of the tropics were always apparent, more so to those who had not previously experienced them. The odoriferous tree cannot be identified.

Section 13

Despite the presumed end of the expedition noted in Section 12, according to the text it continued for another two weeks, nearly half the recorded time. This is either a report of another expedition that has been attached to the previous one, or some error in the extant account, especially regarding the impression that the terminal point had been reached. After having been vague in its details since Section 11, the report becomes more detailed and specific in its last six sections, with topographical and ethnographic comments and the only toponyms and ethnyms since Kerne except for the elusive Aithiopians of Section 11. Whether the text is a linear report or a compilation of more than one expedition, it is clear that the cruise has rounded a major promontory and is approaching the volcanic zone of western Africa. It also seems to have moved into a more populated region, with coastal nightime fires a regular feature. Suetonius Paulinus in the AD 40s was to encounter similar phenomena.[85]

Despite the greater topographical detail, it is difficult to place the features. The high mountains that the expedition sailed around (*peripleusantes*, here using the word in its older sense) indicate a curving coast, and the party may be rounding the place where the African coast ceases running north–south and turns to east–west, with the transition marked by Cape Palmas in eastern Liberia. The "deep gulf" of the ocean cannot precisely be identified, and the perspective of oceanic and coastal features from shipboard can be unreliable. One might think it was the Gulf of Guinea, the concavity where the coast turns from east–west back to north–south, but it is several hundred kilometers across and would hardly be recognizable as a gulf until one crossed it and accurately plotted it. There are bays of various size along much of the African coast, but this one was notable because of its depth (*chasmati*).[86] Yet this is of no help in locating it. Even more enigmatic is the plain on either (or the other) side. The description suggests a river estuary rather than a bay, and since the account has started again from Kerne, as implied in Sections 10–12, it may be back at the Senegal or Gambia, but this means that the rivers are described a second time in totally different terms, and also contradicts the high mountains of Section 12 that the expedition was rounding. Whether the fires were natural or human, or both, is not specified.

Section 14

There would have been a need to water the expedition throughout its course, but this is the only reference to the process. The word used, *hydreusamenoi*, is the standard one, going back to the beginning of Greek literature.[87] Its appearance here is strong support that the "gulf of the ocean" of Section 13 was a river estuary with fresh water rather than a bay of the sea.

The expedition came to the Horn (*keras*) of the West, the first named topographical feature since the Chretes River. The word was originally hydrological—a river inlet—which seems to be its use here, but the descriptive name suggests a promontory, and the word *keras* can also have that meaning, such as in the sense of a "horn" on an animal, although this is not documented before the early fourth century BC.[88] The Horn of the West may be a combination of both these usages: a bay with a promontory sheltering it.

The interpreters had become useful again, although it cannot be taken for granted they were the same ones that had been acquired from the Lixitai

(Section 11). The toponym Horn of the West is a Greek translation of a presumed Punic version of an indigenous name. Since Hanno had no way to determine longitude, the sense of "west," especially on a north–south coast, was arbitrary, and this would suggest that there was something about the feature that made it appear to be toward the west. Moreover, given all the bays noted in the *Periplous*, this one had to be particularly unusual to have its special name, as is the case with the Horn of the South of Section 17.

Any attempt to locate it is purely speculative, and it has been placed all along the coast from Senegal to Nigeria. The best possibility is the Bay of Gorée at modern Dakar, with the volcanic Gorée island located in it. Just to its west is Cap Vert, the westernmost point in Africa. This fits the description, but is far from the mountains and volcanic territory of Section 13.

After the coast turns to run east–west, there are numerous lagoons and lakes, especially along the Ivory Coast, and as far as the delta of the Niger River, which consists of a large number of outlets over a broad expanse. One of these could be the Horn of the West. But the farther one goes east the less viable the concept of "west" becomes, and it is better to locate the Horn of the West on the north–south portion of the coast.

The Carthaginians seem to have been particularly interested in bays with islands in them, since several were mentioned in the *Periplous*. Certainly this would be topography suitable for settlement, and moreover would replicate the harbor of Carthage itself, still visible today. Remains of such harbors are found at a number of places in the Carthaginian territory.[89] Yet there was to be no outpost at the Horn of the West, for even though it seemed peaceful enough in the daytime, at night the locals successfully intimidated the Carthaginians with musical noises, drums, and shouting, all while remaining hidden.

This is the only indication in the *Periplous* that some kind of religious officials accompanied the expedition. The word *manteis* had long been used in Greek literature to mean a religious functionary with prophetic powers, notably at the beginning of the *Iliad*. Such personnel were an essential part of Carthaginian cultic practices.[90] They would have provided guidance to the expedition regarding topographical concerns as well as relations with the locals. Of course, it is nothing but common sense to leave quickly when the inhabitants show hostility but not themselves, yet deferring to the authority of the *manteis* gave the departure a religious underpinning and avoided controversy.

Section 15

Next the expedition entered an active volcanic region. The Gorée Island and Cap Vert area is volcanic, but extinct, and there had been no eruptions since long before Hanno's time. The only other volcanic zone in West Africa south of Morocco is the Cameroon Line, extending southwest from central

Cameroon out to sea, including the chain of islands in the Bight of Biafra (or Bonny) such as São Tomé and Príncipe. Hanno would have needed to reach this region in order to experience fiery incense (*thymiamata*, probably the sulphur odors of a volcanic field), lava flowing into the ocean (as happened in 1922), and hot beach sand. If the Horn of the West was at Gorée Island, the text again has been extremely condensed.

Section 16

The expedition continued through the region of extensive volcanic phenomena. The vulcanism centered on a high mountain that was erupting can be none other than modern Mt. Cameroon (Fako), the highest point (4,095 m.) and the sole active volcano on the West African coast, lying just north of the equator. This is the only place reached by the expedition since Lixos that would seem to be certainly identified. There is no other active volcanic region anywhere close to Hanno's route other than the Canary Islands, but for him to have visited them would require a major rethinking of the evidence for his route.[91]

Nevertheless the volcano would have been a dramatic feature unlike anything seen previously; the imagery that it "seemed to touch the stars" is quite apt and another example of the translator's occasional venture into poetic diction. The interpreters were able to learn its name, which was recorded in Greek as Theon Ochema, or Chariot of the Gods, remarkably similar in concept to one of the contemporary local names, Mongo ma Loba, or Mountain of Heaven.[92] This is the only topographical use of the word *ochema*.[93] The volcano is still active and can be seen from over 100 km. away.

The entry of Theon Ochema into Greek geography became one of the lasting results of the expedition of Hanno, although few from the Greco-Roman world would visit the region. The only possibility is Polybios, who may have reached the area during his expedition along the West African coast after the fall of Carthage in 146 BC. The name does not appear in any of the extant remnants of his report, but he ended up writing a scientific treatise, *On the Inhabited Parts of the Earth Under the Equator*, in which he concluded that the equatorial regions were at high altitude, something that ran counter to contemporary theory. Otherwise it became an exotic toponym in geographical literature, that of one of the largest volcanoes in Africa.[94]

Section 17

Continuing past Theon Ochema, through further volcanic territory, the expedition came to the Horn of the South, the last toponym in the text. It can more easily be located than the Horn of the West. Assuming that the sailing direction is now to the south (the African coast makes its turn back to north–south just north of Mt. Cameroon, at the innermost part of the Bight of

Biafra or Bonny), there are several possibilities for the Horn of the South, all within a fairly small region. One is the Bight of Biafra itself, but this would not easily be defined as three days from Theon Ochema. Another is Cape Lopez, the southern promontory of the bight and the westernmost point on the southern African coast. But since the Horn is in a bay, the more probable locations are Corisco Bay (split between Equatorial Guinea and Gabon) or the Gabon estuary (in northern Gabon). Both these features lie just north of the equator. Wherever the Horn of the South was, according to the text it was the southernmost and final point reached by the expedition.

Section 18

The Horn of the South was a bay with an island within it. This was topographically similar to Kerne and the Horn of the West, leading one to believe that the Carthaginians sought out such features. It was on an island in the Horn of the South that the expedition had its most famous encounter, with the wild people who were locally called the Gorillai. For some reason there was an attempt to capture them, a peculiarity because there had been no such interaction with the populations that they had found previously. This raises the question as to whether the Gorillai were human at all, although the text is explicit in so identifying them (*anthropoi*). They attempted to escape by climbing the cliffs, described by the translator with the exceedingly rare word *kremnobates*. The Carthaginian response, killing and skinning three females, and bringing their hides back to Carthage, implies that they were simian rather than human.[95] The account in Section 18 is unusually vivid and detailed, which may mean nothing more than it was particularly interesting to the translator. The Gorillai are the only ethnic group mentioned by their indigenous name other than the Lixitai.

The Gorillai became assimilated into Greek myth. By Hellenistic times they were thought to be the Gorgons, who had long been located in the far west. They were monstrous sisters, best known through the story of Perseus and Medusa, and lived on the Gorgades Islands, allegedly where Hanno had encountered them, a detail not in the extant *Periplous*. This is the evolution of an exotic remote ethnym into one familiar to the Greco-Roman world, and may have come from Juba's *Wanderings of Hanno*.[96] When the missionary Thomas S. Savage was on the Gabon River in April 1847, he heard reports of a large anthropoid ape, locally called the *engé-ena*, which he named the gorilla after Hanno's wild people.[97] It is unlikely that Hanno saw the elusive gorilla, an animal that does not seem to have been known in classical antiquity, and what he saw was another type of simian, perhaps chimpanzees (*Pan troglodytes*).[98] Not recorded in the *Periplous* is that Hanno put their skins in the Temple of Juno (Astarte) at Carthage, where they were still visible as late as 146 BC.[99]

At this point the text of the *Periplous* comes to an abrupt end, and there is merely a terse final sentence about not sailing any farther because of a lack of provisions. The question of supplies is certainly a relevant one, but Hanno was hardly in a region without food sources. Certainly the expedition turned back at some point and was able to return to Carthage, but if Arrian's report of a 35-day cruise is correct,[100] there are three-and-a-half days missing from the *Periplous*, and there may have been an attempt to sail farther south for a while before giving up, perhaps in order to move beyond the volcanic region. If the original purpose of the expedition was to encircle Africa, the realization that the coast was now going to the south for an indefinite distance could have been the final blow. There was a persistent belief, at least in the Greek world, that Africa was long and narrow, and that its east–west coast which Hanno had been sailing along ran straight across to East Africa: this view was still persistent in the late second century BC.[101] The sharp turn south at the Bight of Biafra may have made the idea appear impossible, and this, coupled with other concerns, put an end to the expedition. As Arrian put it, probably quoting from Juba's *Wanderings of Hanno*:

> When he turned to the south, he happened upon all kinds of unim-aginable problems: a lack of water, burning heat, and streams of fire flowing into the earth.

The lack of water is the least credible excuse advanced, but the reasons cited by Arrian for the end of the cruise were presumably what Hanno reported at Carthage to those who had no idea of what he had experienced. Hanno was not the only one to blame extraordinary circumstances for terminating a West African expedition: at about the same time the cruise of Euthymenes of Massalia was brought to an end, probably because of a failure to understand the tides—they were little known in the Mediterranean world—and a generation later Sataspes of Persia claimed that his ship became stuck, perhaps another example of tidal phenomena.[102] Hanno may also have found the tides baffling, and was probably aware that adverse winds would make the return more difficult the farther he went.

Conclusion

Hanno went a greater distance south along the West African coast than anyone in antiquity except the elusive Phoenicians who had circumnavigated the continent a century previously. The purpose of the expedition cannot easily be defined.[103] It was originally intended to establish Carthaginian out-posts in West Africa, as explicitly stated at the beginning of the *Periplous*. But no settlements were created after Lixos, which was near the beginning of the cruise, and the project diversified and evolved into true exploration, although there is no recorded actual agenda for this activity. There may have been an interest in the sources of precious metals, but this is speculative.[104] Carthaginian trading outposts became a common feature along this coast—allegedly there were as many as 300 at one time[105]—yet these were hardly all established by Hanno. Kerne became the dominant site, but it and the others faded as Carthaginian power declined. Later knowledge of Hanno and his expedition was surprisingly minimal despite the extant text. Scattered sporadic knowledge about Hanno entered the Greek intellectual horizon, yet in a contradictory and confused fashion. Those who knew the most about Hanno were the Massalians, with their own interests beyond the Pillars of Herakles, but they did not share their data with others since all reconnaissances that had a commercial aspect—including their own–were trade secrets. The Romans attempted to learn more about the west coast of Africa by sending Polybios into the region in 146 BC, but he found little other than decline and abandonment. By the following century it was easy to be dismissive of Hanno, a point of view that unfortunately is still promoted.[106] Nevertheless Hanno was one of the great explorers of antiquity, and his text remained the most detailed account of the West African coast for nearly 2,000 years.

Notes

1   *On Marvellous Things Heard* 37, in the Aristotelian corpus.
2   Herodotus 4.196; Pseudo-Skylax 112; Palaiphatos, *On Unbelievable Tales* 31.
3   Blomqvist, *Date* 18–51.
4   Herodotus 4.196.
5   Pseudo-Skylax 112; Graham Shipley, *Pseudo-Skylax's Periplous: The Circumnavigation of the Inhabited World* (Bristol 2011) 6–8.
6   Eratosthenes, *Geography* F13.
7   Polybios 34.15.9; Pliny, *Natural History* 5.8–11; 6.199; 18.22–3.
8   Athenaios 3.83b; Pliny, *Natural History* 6.200; Duane W. Roller, *Scholarly Kings: The Writings of Juba II of Mauretania, Archelaos of Kappadokia, Herod the Great, and the Emperor Claudius* (Chicago 2004) 43–6.
9   Pomponius Mela 3.90–5; Pliny, *Natural History* 2.169, 5.8, 6.200; see also Arrian, *Indike* 43.11–12.
10 Carolina López-Ruiz, "Phoenician Literature," in *The Oxford Handbook of the Phoenician and Punic Mediterranean* (ed. Carolina López-Ruiz and Brian R. Doak, Oxford 2019) 256–69.
11 Roller, *Through the Pillars* 31–3.
12 Casson, *Ships* 273–8; Blomqvist, *Date* 10–11.
13 Plato, *Timaios* 24e–25d; Theopompos (*FGrHist* #115) F75c; Hekataios of Abdera (*FGrHist* #264) F7–14; Euhemeros of Messene (*FGrHist* #63) F2–3.
14 Infra.
15 For vestiges of Punic constructions, see Stanislav Segert, "Phoenician Background of Hanno's Periplus," *MUSJ* 45 (1969) 501–18.
16 Blomqvist, *Date* 46–7.
17 Homer, *Iliad* 1.423, 23.206; *Odyssey* 1.22–5.
18 Ephorus F30a = Strabo, *Geography* 1.2.28.
19 Herodotus 1.166, 4.152.
20 Pliny, *Natural History* 18.22–3.
21 Cornelius Nepos, *Hannibal* 13; Hannibal (*FGrHist* #181) F1.
22 Hanno the Carthaginian, *Periplus* (ed. Al. N. Oikonomides and M. C. J. Miller, Chicago 1995).

23 C. Müller, *GGM* vol. 1, pp. 1–14.
24 M. Euzennat, "Pour une lecture marocaine du *Périple* d'Hannon," *BCTH* 12–14b (1976–8) 243–6.
25 M. Sznycer, "Carthage et la civilisation punique," in *Rome et la conquête du monde méditerranéen* (ed. Claude Nicolet, Paris 1977–8) 565–76; Werner Huss, "Carthage [IA]," *BNP* 2 (2003) 1130.
26 There are over 30 known Carthaginians with the name Hanno, but the explorer cannot be related to any of them, or any other of his compatriots. He is no. 3 in Klaus Geus, *Prosopographie der literarisch bezeugten Karthager* (Leuven 1994) 98–105.
27 Hanno, *Periplous* 1, 8, 18; Herodotus 1.166 etc.; Sophokles F602 (*Triptolemos*).
28 Homer, *Odyssey* 4.85; see also Herodotus 4.42.
29 Herodotus 2.44.
30 Homer, *Odyssey* 1.51–4; Aeschylus, *Prometheus Bound* 349–52; Herodotus 2.33 etc.; Adolf Schulten, *Iberische Landeskunde: Geographie des antiken Spanien* (Strasbourg 1955–7) 399–404.
31 Paolo Xella, "Religion," in *The Oxford Handbook of the Phoenician and Punic Mediterranean* (ed. Carolina López-Ruiz and Brian R. Doak, Oxford 2019) 282–5.
32 Sznycer, "Carthage" 576–85.
33 Polybios 3.33.15; see also Pliny, *Natural History* 5.24; Avienus, *Ora maritima* 421.
34 Homer, *Iliad* 16.169–70; Pindar, *Pythian* 4.245; Herodotus 1.152 etc.; Casson, *Ships* 44; René Rebuffat, "Les pentécontores d'Hannon," *Karthago* 23 (1995) 20–30.
35 J.-C. Demerliac and J. Meirat, *Hannon et l'empire punique* (Paris 1983) 65–7.
36 Herodotus 4.42, 44.
37 Homer, *Odyssey* 9.9.
38 Herodotus 3.138, 4.152.
39 Pliny, *Natural History* 2.169.
40 Pliny, *Natural History* 19.3–4.
41 Strabo, *Geography* 3.5.5–6; Duane W. Roller, *A Historical and Topographical Guide to the Geography of Strabo* (Cambridge 2018) 167–9.
42 Diodoros 4.17.4, 4.27.3; Pomponius Mela 1.26; Pliny, *Natural History* 5.2; M. Euzennat, "Tingi," *PECS* 923; Michel Ponsich, "Tanger antique," *ANRW* 2.10.2 (1982) 798–800.
43 Herodotus 4.162.
44 Herodotus 2.32, 42; see also Pseudo-Skylax 112; Shipley, *Pseudo-Skylax's Periplous* 201–3.
45 Pseudo-Skylax 112; Xella, "Religion" 276–7.
46 The discussions are summarized in Roller, *Through the Pillars* 35.
47 Strabo, *Geography* 17.3.5; Suzanne Amigues, "Végétaux étranges ou remarquables du Maroc antique d'après Strabon et Pline l'Ancien," *AntAfr* 38–9 (2002–3) 46–8.
48 Juba (*FGrHist* #275) F47–54, 58; H. H. Scullard, *The Elephant in the Greek and Roman World* (Cambridge 1974) 30–1.
49 Herodotus 1.96, 2.154.
50 Strabo, *Geography* 17.3.3, 8.
51 Jacques Ramin, *Le Périple d'Hannon* (*BAR Supplementary Series* 3, 1976) 82.
52 Ephorus (*FGrHist* #70) F53.
53 Hekataios (*FGrHist* #1) F357.

54 Homer, *Odyssey* 4.84; Strabo, *Geography* 16.4.27, Pliny, *Natural History* 5.2; M. Euzennat, "Jibila," *PECS* 426.

55 Hekataios of Miletos (*FGrHist* #1) F355; Pliny, *Natural History* 5.3–4; Jehan Desanges, "Lixos dans les sources littéraires grecques et latines," in *Actes du colloque de Larache, 8–11 novembre 1989* (Rome 1992) 1–6; Euzennat, "Lixus," *PECS* 521.

56 Duane W. Roller, *The World of Juba II and Kleopatra Selene: Royal Scholarship on Rome's African Frontier* (London 2003) 133–5.

57 Strabo, *Geography* 17.3.2.

58 Sophokles, *Women of Trachis* 762; Euripides, *Bakchai* 677. Their cattle were perhaps zebu: see H. R. Palmer, "The Lixitae of Hanno," *JRAS* 27 (1927) 7–15.

59 Pausanias 1.33.5.

60 Supra, pp. 17–18.

61 Herodotus 4.183; Pliny, *Natural History* 6.169; G. W. Murray, "Trogodytica: The Red Sea Litoral in Ptolemaic Times," *GJ* 133 (1967) 24–33.

62 Hesiod, *Works and Days* 715; it was also used by the dramatists.

63 Serge Lancel, *Carthage: A History* (tr. Antonia Nevill, Oxford 1995) 105–6.

64 Herodotus 4.43; an anecdote obtained from Carthaginian sources.

65 Pliny, *Natural History* 5.14–15.

66 Homer, *Odyssey* 3.269–72; Herodotus 4.18.

67 Herodotus 2.125, 154.

68 René Rebuffat, "Les nomades de Lixus," *BCTH* 18 (1982) 78–80.

69 *GGM* vol. 1, p. 7; Jérôme Carcopino, *Le Maroc antique* (tenth edition, Paris 1948) 131–2.

70 Pseudo-Skylax 112.

71 Casson, *Ships* 284–7. The sail of 6.2 knots was from Corinth to Puteoli in the first century AD (Philostratos, *Life of Apollonios* 7.10).

72 Shipley, *Pseudo-Skylax's Periplous* 208.

73 Pliny, *Natural History* 6.201; Roller, *World of Juba* 115–17.

74 Roller, *Through the Pillars* 37.

75 Pseudo-Skylax 112; Polybios 34.15.9; Strabo, *Geography* 1.3.2; Ptolemy, *Geographical Guide* 4.6.33.

76 Herodotus 4.196.

77 Blomqvist, *Date* 26.

78 Aristotle, *Meteorologika* 1.13.350b; see also *GGM* vol. 1, p. 8.

79 Homer, *Iliad* 16.324, when Thrasymedes crushed the bones of his opponent; Herodotus 8.90.

80 Herodotus 2.68–70; Kenneth F. Kitchell, Jr., *Animals in the Ancient World From A to Z* (London 2014) 37–8.

81 Herodotus 3.71; Kitchell, *Animals* 87–8.

82 Euzennat, "Pour une lecture" 243–6.

83 For the scattered sources on Euthymenes, see Roller, *Through the Pillars* 15–19.

84 Francis Lacroix, "Les langues," in *Histoire générale de l'Afrique noire* 1 (ed. Hubert Deschamps, Paris 1970) 76–7.

85 Pliny, *Natural History* 5.7.

86 This word was not used elsewhere in the extant *Periplous*, and tends to be poetic. It has a sense of immense depth, not unlike its modern derivative. Hesiod used it to describe Tartaros (*Theogony* 740), and Herodotus for the deep gorge of the Lykos River in Phrygia (7.30).

87 Homer, *Odyssey* 10.105.

88 Hesiod, *Theogony* 789; Xenophon, *Anabasis* 5.6.7. It may also be an indigenous name, "kor", for "river": Carcopino, *Maroc antique* 131–2.

89 See also Sections 9, 18. For Carthaginian harbors, see Strabo, *Geography* 17.3.14–15, and Jerker Blomqvist, "Reflections of Carthaginian Commercial Activity in *Hanno's Periplus*," *OrSue* 33–5 (1984–6) 55–7.

90 Homer, *Iliad* 1.62; Xella, "Religion" 285–7.

91 Alfredo Mederos Martín, "La exploración del litoral atlántico norteafricano según el periplo de Hannón de Cartago," *Gerión* 33 (2015) 29–31.

92 D. B. Harden, "The Phoenicians on the West Coast of Africa," *Antiquity* 22 (1948) 145–6.

93 For other citations, generally meaning "support" or "chariot," see Herodotus 5.21; Euripides, *Trojan Women* 884; Roller, *Through the Pillars* 40.

94 Polybios 34.1.7, 16; Geminos 16.32; Pomponius Mela 3.94; Pliny, *Natural History* 2.238, 5.10; Ptolemy, *Geographical Guide* 4.6.9, 16.

95 Blomqvist, *Date* 35; F. E. Romer, *Pomponius Mela's Description of the World* (Ann Arbor 1998) 127.

96 Hesiod, *Theogony* 270–83; Pliny, *Natural History* 6.200; Athenaios 3.83d.

97 Thomas S. Savage, "Notice of the External Characteristics and Habits of Troglodytes Gorilla, a New Species of Orang From the Gaboon River," *Boston Journal of Natural History* 5 (1847) 417–43. This was probably the western gorilla, *Gorilla gorilla*, although Savage named it *Troglodytes gorilla*.

98 Kitchell, *Animals* 77–8.

99 Pliny, *Natural History* 6.200.

100 Arrian, *Indika* 43.11–12.

101 Strabo, *Geography* 2.3.4.

102 Herodotus 4.43; Roller, *Through the Pillars* 15–20.

103 Ramin, *Périple* 93–7.

104 Roller, *Through the Pillars* 41–2.

105 Strabo, *Geography* 17.3.3.

106 For examples of the negative view, see Roller, *Through the Pillars* 31–2.

3

THE *PERIODOS* DEDICATED TO
KING NIKOMEDES

In the late second century BC an author whose name is no longer known wrote a poetic description of the circuit of the Mediterranean system, from the Pillars of Herakles at the west circling around in a clockwise fashion, and presumably returning across the northern coast of Africa to the Pillars.[1]

The work is indebted to the *periplous* format, but is styled a *periodos*, a circuit of the earth, and there are numerous divergences from the coasts. Like Hanno's report, it ended up in the corpus of the *Minor Greek Geographers*, the very last item in the late thirteenth-century manuscript known as *Codex Parisinus graecus supplementi* 443. Yet the end of this manuscript vanished before the sixteenth century, and with it a large portion of the text of the *Periodos*, as well as the name of the author, which was normally placed at the end of the relevant text.

Only the first 747 lines of the text survive, ending with the author's description of Mt. Haimos in Thrace. Another 278 lines, describing the circuit of the Black Sea, can be extracted from metrical passages embedded in a document called the *Periplous of the Euxine Sea*, a compilation probably composed no earlier than the latter part of the sixth century AD.[2] If the *Periodos* continued at the same level of detail south through western Asia Minor and the Levant, and then across northern Africa back to the Pillars of Herakles, probably at least half of the text is lost. The work is dedicated to an unspecified King Nikomedes, one of the Hellenistic rulers of Bithynia in northwestern Asia Minor.

The author of the *Periodos*

The lack of an author's name has resulted in many efforts to supply one. Given that nearly 250 writers on geography are known from classical antiquity, practically none of whom is extant, this is probably an exercise in futility.[3] In 1630 Lucas Holsten attributed the *Periodos* to Skymnos of Chios, a barely preserved geographer of the early second century BC, who was the son of Apelles of Chios and became a *proxenos* at Delphi in 185/184 BC. Only a

     DOI: 10.4324/9781003030379-4

Map 3.1 Major places cited in *The* Periodos *Dedicated to King Nikomedes.*
Source: Map by E. Rodriguez

few fragments of his work remain.[4] August Meineke realized in 1846 that this attribution was not possible, and proposed calling the author Pseudo-Skymnos. Unfortunately this name remains in use nearly two centuries later, even though the text has nothing to do with Skymnos himself.[5] The text will be referred to herein as the *King Nikomedes Periodos* or merely the *Periodos*.

The question remains as to who the actual author was. There have been a number of suggestions, none of them definitive. Perhaps the most reasonable is Pausanias of Damascus, an obscure geographer cited by the tenth century Byzantine emperor Constantine VII.[6] Yet the association is ephemeral at best, and assumes that a Syrian Greek ended up at the Bithynian court in the late second century BC. But Constantine seems to have had access to the collection of the *Minor Greek Geographers*, perhaps without the loss of the end of the *King Nikomedes Periodos*.

Although there are many other possibilities for authorship, two additional ones deserve special consideration. An elusive reference (without name) to Apollodoros of Athens near the beginning of the poem (lines 16–44) raises the idea that he was actually the author. But there seems no good reason that Apollodoros, one of the major scholars of the second century BC and famous for a variety of influential writings, including his definitive *Chronika*, would have any reason to be so coy, especially when dedicating a work to Hellenistic royalty.[7] The geographer Semos of Delos, who wrote two *periodoi*, has also been proposed.[8] But these possibilities lack any evidence that would make them credible, and the fact remains that the name of the author will probably never be known and may not have even been cited elsewhere in extant literature. Yet an interesting point is that when the author drew attention to his personal experiences (lines 128–36), the emphasis is on the world west of the Greek peninsula (the Adriatic and Ionian Seas, Tyrrhenia, Sicily, Libya, and Carthage). Otherwise only Greece and "the settlements lying around Asia" are mentioned, raising the possibility that he came from a more western location: one might even suggest Carthage or North Africa, since that city is the only one mentioned by name in this context.

King Nikomedes

Even if the identity of the author of the *Periodos* is not known, and probably never will be, the context of the composition is no mystery. In the second line the author reported that he dedicated his work to a certain king Nikomedes, whose name also appears in the titular heading of the text, which, however, may have been added by a later compiler. Although the name of the king does not appear again in the extant manuscript, there is little difficulty in understanding the environment in which the treatise was produced.

There is only one Hellenistic royal family that made use of the name Nikomedes, the rulers of the minor state of Bithynia in northwest Asia Minor. It had been an independent kingdom since the second half of the fifth century BC, and managed to survive as such into late Hellenistic times. Four kings named Nikomedes ruled it from the early third century BC to 74 BC.[9] The first was too early for the patron of the author of the *Periodos*, since later sources are mentioned in the text and his reign was well before any contact between Bithynia and Rome, mentioned in lines 231–5. When Nikomedes I died, around 255 BC, Rome was still an unknown factor in the politics of Asia Minor.

It was only in 169 BC that Bithynia sought relations with Rome, when King Prousias II sent an embassy to the city.[10] The *Periodos* must date to after this time, and after Prousias' son Nikomedes II came to the throne in 149 BC, having been involved in a successful plot to kill his father. He had already visited Rome, and later assisted the Romans during the unstable years after the end of the Pergamene kingdom in 133 BC.[11] He survived into the 120s BC, and has been suggested as a candidate for the dedicatee of the *Periodos*.

Nikomedes II was succeeded by his son Nikomedes III, who ruled until around 94 BC. His reign was one of increasing Roman involvement in Asia Minor, as well as the rise of the most important regional dynast of the era, Mithridates VI of Pontos. Nikomedes spent most of his reign in various degrees of entanglement with both. He was known for his generosity, and took the surname Euergetes ("Benefactor").[12] Although he experienced economic difficulties and increasing issues with Rome, he is the other candidate for the patron of the author of the *Periodos*, indeed the only reasonable one, and the diffidence of previous commentators about the particular king in question is not necessary.[13]

To be sure, there was one additional Nikomedes who ruled Bithynia, the fourth and the son of Nikomedes III, whose reign lasted from about 94 BC to 74 BC, when he willed his kingdom to Rome.[14] Yet he was weak and undistinguished, and was pressed even more by both Rome and Mithridates VI. His rule was generally unstable, and in fact part it was spent in exile. He was hardly the patron of culture mentioned in the *Periodos* (lines 50–65), and can easily be eliminated as the king in question.

Thus the patron of the *Periodos* was Nikomedes III. This is shown indisputably by the reference to the king's father (line 56). Nikomedes II came to the throne after his father, Prousias II, was killed in 149 BC.[15] Any comment, however vague, on the relationship between those two kings would have been immensely inappropriate, since the son was implicated in his father's death. Nikomedes III, on the other hand, succeeded his father Nikomedes II peacefully in the early 120s BC, a few years after the end of the Pergamene kingdom (lines 16–18). Thus, in conclusion, the *Periodos* was written by an unknown author in probably the last quarter of the second century BC and dedicated to King Nikomedes III of Bithynia.

The Poem of the *Periodos*

The author of the *Periodos* was inspired by Apollodoros of Athens in using iambic trimeter for his composition. This was a meter that had a long and distinguished history in Greek literature. It originated in the seventh century BC, perhaps with Archilochos, and became a standard meter of Greek drama and comedy. Apollodoros lived during the reign of Attalos II of Pergamon (159–138 BC), so was an older contemporary of the author of the *Periodos* (lines 45–9), perhaps even his teacher.[16] He was probably the first historical writer to use the meter; his motivation was that the metrical form provided greater clarity and structure than prose, as well as a certain stylistic grace (lines 33–44). This was a standard view of the virtues of metrical writing over prose, as Dionysios of Halikarnassos was to point out.[17] Presumably these reasons were also relevant to the *Periodos*, although the modern reader cannot fail to notice certain place holders (such as the repeated use of *logos*) to fill out lines, as well as idiosyncratic but metrically valid forms of toponyms and ethnyms. Moreover, the poetry is often particularly close to prose in form, perhaps a definite attempt on the part of the author to straddle the two genres and create a prosaic poetic style. This may also be a result of his reliance on prose sources.[18]

The text of the *Periodos* opens with an address to King Nikomedes, stressing the value of comic meter in explaining matters briefly and distinctly, and including praise of Apollodoros, although he was not mentioned by name (lines 1–64). Next there is a summary of the types of material contained in the work and a list of the primary sources (lines 109–36). The actual geographical narrative begins at the Pillars of Herakles (lines 137–69), but digresses to discuss the ethnic distribution of the inhabited world (lines 170–82) before returning to a description of the western Mediterranean, Italy, Sicily, and the Adriatic (lines 183–443).

The account then considers the Greek peninsula, the region discussed in the greatest detail in the extant text (lines 444–663). The final surviving lines examine the upper Aegean and Thrace as far as the Black Sea (lines 664–747). An additional 278 lines, describing the circuit of the Black Sea, can be restored from quotations in the late-antique *Periplous of the Euxine Sea*.

The author mentioned 14 predecessors in the extant text. Two of these are not truly sources, but the teachers of Apollodoros: Diogenes the Stoic and Aristarchos of Samothrace, both active in the second century BC. Although it is unlikely that either of these were directly used by the author of the *Periodos*, mentioning them allowed him to place himself within the literary and scholarly tradition of the late Hellenistic world, including Apollodoros, who was not named.

The author also listed his direct sources (lines 109–36). The text is highly corrupt, and although nine names appear, at least one and probably others have dropped out. The earliest cited is Herodotus, and the remainder extend from the fourth into the second century BC. The structure of the list is not totally at random: it begins, reasonably, with Eratosthenes of Kyrene, the inventor of the discipline of geography, from the second half of the third century BC, to whom the author stated his great indebtedness, and ends with Herodotus, the originator of historical research. Between these two seminal scholars are seven extant names. Four of these do not appear in the preserved text (Dionysios, Kleon, Timosthenes, and Kallisthenes), but this is of little significance given that not all of the *Periodos* survives. Three additional ones are in the extant text (Hekataios of Teos, Theopompos, and Antiphanes) but not in this list, yet again its fragmentary nature means that they may have been cited in the partial and missing lines between Timosthenes and Timaios. In addition, the author tells his readers of his own wide-ranging autoptic experience, from Italy and Carthage in the west to the Greek world and Asia Minor (lines 128–36).

Implicit also in the source material for the *Periodos* is a genre that might be called "foundation literature." These were accounts of city foundations that recorded the history of the establishment of towns, from their presumed mythological origins into historical times. The earliest known is perhaps the work of Xenophanes of Kolophon in the sixth century BC. He is better remembered for his innovative philosophical and religious views, but he also wrote *The Foundation of Kolophon* and *The Settlement of Elea in Italy.*[19] Others followed in this genre, and like Xenophanes they usually wrote about cities with which they were associated, such as Kadmos of Miletos and Ion of Chios. Yet Hellanikos of Lesbos produced a more general treatise, *The Foundation of Peoples and Cities.*[20] This emphasis on city foundations and ethnicity became part of the Greek self-image, given the establishment of so many new Greek cities outside the traditional heartland from the eighth century BC onward. The author of the *Periodos* made wide use of such material, although inevitably synthesized through later authors such as Ephorus and Timaios. But the accounts in the *Periodos* are varied in their detail, with some places having little or no material beyond toponyms and ethnyms, and others with a fuller discussion. Nevertheless the *Periodos* remains a primary source for foundation details of many Greek cities.[21]

The translation

The following translation is based on the edition of Didier Marcotte, supplemented by the restoration of lines 748–1026 as suggested by Aubrey Diller.[22] The text is in poor shape; in addition to the loss of everything after line 747 in the manuscript, as well as the author's name, there are missing lines and gaps throughout, and in one place (lines 390–401) the narrative has become jumbled. Yet an unintended consequence of the metrical structure is that restorations are easier than in a prose text, and even if the surviving text is not grammatically pure, its sense is often clear. The manuscript has the putative title of the treatise, *Periodos of the Earth in Comic Meter For King Nikomedes*, but this may have been added by one of the later compilers.

The text

(1) What is most completely necessary for comedy, most divine King Nikomedes, is that it tell everything briefly and distinctly, and that it be alluring in every way to one who is a wise judge. Thus, having demonstrated the persuasiveness of the style, I was eager to have an interview with you about it and also a brief discussion. The work that has been produced, although limited, is intended to be advantageous to all, providing a common benefit to anyone who wishes through you to have a love of learning. Desiring, then, first to place the reason for the entire composition clearly before you, I request that a brief word be set forth as an introduction, since I have decided, in the Lakonikan manner, to speak of great matters in the shortest way.

(16) These things are what I write about. For the benefit of the kings of Pergamon—whose reputation remains with us for all time, even after their death—a certain genuine Attic scholar,

who had become a listener of Diogenes the Stoic, and who also had studied for a long time with Aristarchos, composed a chronography that went from the Trojan capture down to the present time. He laid out in a definitive manner 40 in addition to a thousand years, enumerating the capture of cities, the movement of military camps, the removal of peoples, barbarian expeditions, attacks and movements by naval forces, the location of festivals, alliances, treaties, battles, the deeds of kings, the lives of distinguished men, fugitives, expeditions, and the abolition of tyrannies: a summary of everything that had been described indiscriminately. He chose to set this in metrical form, that of comedy, for the sake of clarity, seeing that it would be easily memorized. He took an image portrayed from life: if someone wished to pick up and carry a loose quantity of wood, he would not be able to lay hold of it readily, but it would be easier if it were tied together. Thus one cannot learn loose phrases quickly, but if they are encompassed in meter they can be grasped unerringly and faithfully, for there is a gracefulness running through it when history and language are woven metrically.

(45) That person, having gathered together the main points of chronology, submitted them to the favor of King Philadelphos. They went through the entire *oikoumene*, imparting undying repute on Attalos [II], who received the dedication of the work. But I, hearing that you alone of contemporary kings demonstrate royal excellence, wished to prove it to myself by coming to you in order to see what a king is, and so that I can report it back to others myself. Thus I chose as advisor for the project the one who once assisted in normalizing the matters of your kingdom for your father, as we hear, and who is truly honored by you, King, in everything: I mean Apollo of Didyme, the giver of oracles and leader of the Muses. Trusting almost completely in him, I came to your hearth at your command, which you had proclaimed was almost in common for lovers of learning. Let the god lay hold of my choice!

(65) Using what some have recorded in a scattered manner, I have written for you a summary of the settlements, the foundation of cities, and the places that are navigable and passable over almost the entire earth. Among them, I will expound succinctly what is conspicuous and clear, the important points, but what is not clearly known, I will present in precise detail. Thus, King, you will have the entire *oikoumene* marked out concisely: the characteristics and courses of the great rivers, the position of the two continents according to their parts, which cities in each are Hellene, who founded them, at what time they were inhabited, those which are of the same ethnicity, and those which are

indigenous, which types of barbarians are adjacent and which are called mixed, which are nomadic and which are civilized, which are most inhospitable in their customs or most barbaric in their actions, which ethnicities are the greatest and most populous, what laws and livelihood each has, which emporia are the most successful, the position of all the islands off Europe and those lying near Asia, and the city foundations recorded in them. Simply, it is a description of all places and the entire circuit in a few lines. Whoever listens will not only have pleasure but also take away the useful benefits of knowledge: if nothing else, they say, where exactly he is on the earth, in what places his fatherland lies, which inhabitants it formerly belonged to, and with which cities it attributes kinship. To state it briefly: without taking up what the myths call the Wanderings of Odysseus but remaining happily in his own pastures, someone will not only learn about differences in human life but the towns and customs of everyone. The composition, with you as its most distinguished leader and favorable guardian, will move through an attentive childhood into life, and will announce your fame, King, to all as it is sent from place to place, revealing your honor to those far away.

(109) I will now begin my composition by setting forth the writers whom I am using to move my historical work toward credibility. I am most persuaded by the one who has written most attentively about geography, and the *klimata* and the figures, Eratosthenes, as well as Ephorus [T32], and the one who recounted about foundations in five books, Dionysios the Chalkidean, as well as Demetrios, the writer from Kallatis [T5], the Sicilian Kleon, and Timosthenes [F5], who positioned...and the citizen [two lines illegible] places...also following Kallisthenes [T34]...and Timaios [T28], a Sicilian man from Tauromenion, as well as what was composed by Herodotus. There is also what I have personally and diligently scrutinized, bringing autoptic reliability to it; having observed not only Hellas and the settlements lying around Asia, I have become skilled in those around the Adria and the Ionian Seas. I have gone to the boundaries of Tyrrhenia, Sikelia, and toward the west, and to most of Libya and Karchedon.

(137) Bringing together many things, I will begin the matter at hand. First I will put in order the places around Europe. The mouth of the Atlantic Ocean, they say, is 120 stadia. The nearby lands enclosing it are Libya and Europe. Islands lie on either side of these, at about 30 stadia from each other, which some call the Pillars of Herakles. Near one of these is a Massaliote city called Mainake, whose location is that of the farthest of all the Hellenic cities in Europe. Going around the cape opposite to

the setting sun, there is a sail of one day, and then there is the island called Erytheia, which is quite short in its length, and has herds of cattle, wild and domesticated, that resemble Egyptian bulls, as well as the Thesprotian ones in Epeiros. They say that the inhabitants are Western Aithiopians, who made a settlement. Nearby there is a city, Gadeira…which took as settlers Tyrian merchants, and where it is reported that great sea monsters are produced. After this, upon completing a sail of two days, there is a most successful emporium, called Tartessos, a famous city, to which tin, gold, and a large amount of copper is washed down from Keltike. The territory called Keltike is next, lying as far as Sardo in the sea, the largest population toward the sunset.

(170) The Indians live almost everywhere within the sunrises, but toward the midday are the Aithiopians, who lie near the breath of the south wind. The Kelts extend from the *zephyros* as far as the summer sunset, and the Skythians are toward the *boreas*. The Indians live between the summer and winter sunrises, but by contrast the Kelts are between the equinoctial <and summer> sunset, as it is said. These four peoples are equal in their populations and in the number of their inhabitants, but the territory of the Aithiopians is somewhat larger, as is that of the Skythians, although for both a large amount is deserted because the former have more that is fiery and the latter that which is damp.

(183) The Kelts make use of Hellenic customs, and have a close relationship with Hellas because of receiving them as guests. They conduct their assemblies with music, esteeming it because of its civilizing effect. The farthest place among them is the so-called Column of Baria [?], which is exceedingly high and is a promontory extending into the billowy ocean. In places near the column live the Kelts who are called the Farthest, as well as the Enetians and those near the Adria who are within the Istros [?]. They say that from here the Istros takes the beginning of its flow.

(196) The Libyphoenicians live along the open Sardoan Sea and established a settlement from Karchedon. Next, it is believed, the Tartessians lived there, and then the neighboring Iberians. Above these places are the Bebrykes. Below them and along the coast are the Ligyes and the Hellenic cities that the Massaliote Phokaians settled. First is Emporion, with Rhode second, which was founded by those who formerly controlled the sea, the Rhodians. After them the Phokaian founders of Massalia came to Iberia and took Agathe and Rhodanousia, beside which flows the great Rhodanos River. Then there is Massalia, an especially great city and Phokaian settlement, which, it is said, they founded in Ligystine 120 years before the battle that occurred at Salamis. This is what Timaios [F71] records about the foundation. Next after it is Tauroeis, with the city of Olbia nearby, and Antipolis, the farthest one.

(217) After Ligystine are the Pelasgians, who settled there from
Hellas at an early time, sharing the land in common with the
Tyrrhenians. The Lydian Tyrrhenos, the son of Atys, founded
Tyrrhenia, having previously come to the Ombrikoi. There are
islands in the strait and in the open sea: Kyrnos, and Sardo,
which is said to be the largest island after Sikelia. There are also
those formerly called the Seirenides, and the islands of Kirke.
The Ombrikoi are above the Pelasgians... <the Latins> who
were founded by Latinos, the offspring of Kirke and Odysseus,
as well as the Ausones, who have a place in the interior and
seem to have been unified by Auson, the offspring of Odysseus
and Kalypso. Among these populations is the city of Rome,
whose name rivals its power, a common star of the entire
oikoumene, in Latine. They say it was founded by Romulus, who
gave it its name.

(236) After the Latins there is a city among the Opikoi and near
the so-called Lake Aornos, Kyme, which was originally settled
by the Chalkideans, and then by the Aiolians. Here a certain
Kerberion is shown, an underground oracle, and they say that
Odysseus came to it after returning from Kirke. It was from
Kyme—which is located next to Aornos—that Neapolis was
founded, according to an oracle. The Saunitai live near them,
who were after the Ausones. After them, lying in the interior,
live the Leukanoi and the Kampanoi. Adjoining them, moreover, are the Oinotrioi, as far as what is named Poseidonias,
which they say was first settled by the Sybarites. Then there is
Elea, a city of the Massaliote Phokaians, which the Phokaians
founded when fleeing the Persian situation. The city of Phokaia,
situated in Asia, was extremely abundant in men.

(254) In the Tyrrhenic Strait there are seven islets not far from
Sikelia, which they call the Islands of Aiolos. One of them is
reasonably called Hiera, since burning fires appear from it, visible to all from a number of stadia away, with fiery anvils thrown
up to a height, the result of the work of the crashing of iron
hammers. One of them has a Dorian settlement named Lipara
which is kindred to Knidos.

(264) Next there is the extremely prosperous island of Sikelia.
They say that formerly Iberian barbarian multitudes with different languages divided it up. Because of the several-sided
nature of the territory, it was called Trinakria by the Iberians,
but in time it was renamed Sikelia, during the domination of
Sikelos. Then it had Hellenic cities, as they say, in the tenth
generation after the Trojan matter, when Theokles obtained a
fleet from the Chalkideans (although he was ethnically from
Athens), and, as is said, there were Dorian and then Ionian
inhabitants.

(276) When civil discord occurred among them, the Chalkideans founded Naxos and the Megarians Hybla, and the Dorians took possession of Epizephyrion in Italia. Archias the Corinthian along with the Dorians took these and founded a place named after a bordering lake, today called Syrakousai by them. After this, the city of Leontine—which was located opposite Rhegion and on the Sikelian Strait—was established as a settlement from Naxos, and Zankle, Katane, and Kallipolis were founded. Then from these the two cities called Euboia and Mylai were settled, and then Himera and Tauromenion. All these were Chalkidean cities. It is necessary to speak of the Dorians again: the Megarians founded Selinous and the Geloans Akragas. Messene was by Ionians from Samos, and the Syrakosioi what was called Kamarina, which they took down to its foundations again after it had been settled for 46 years. These are the Hellenic cities, but the remaining settlements are barbarian, places that the Karchedonians fortified.

(300) Italia adjoins Oinotria, and formerly contained mixed barbarians, taking its name from Italos, who held power there. Later it was called Great Hellas Toward the West because of its settlements. Indeed it has Hellenic coastal cities: first Terina, which had previously been settled by the Krotoniates, and then nearby Hipponion and Medma, which the Lokrians founded, and then the Rheginoi and the city of Rhegion, where there is the shortest crossing by ship to Sikelia. It seems that Rhegion was a Chalkidean settlement. The so-called Epizephyrian Lokrians are nearby, whom they say were the first to use written laws, which Zaleukos seems to have set down. They are settlers from Opountian Lokris, although some say from the Lokrians in Ozolai.

(318) First after these is Kaulonia, which was settled from Kroton and named for the hollow [*aulon*] that is near the city, but at a later time the name was changed to Kaulonia. Then there is the formerly successful and populated city of Kroton, which seems to have been founded by Myskelos the Achaian. After Kroton are Pandosia and Thourioi, which border Metapontion. They say that all these cities were founded by Achaians coming from the Peloponnesos. Then there is the largest one in Italia, Taras, called after a certain hero, Taras, a Lakedaimonian settler. It is a fortunate city. The Partheniai once founded it, and it is well situated, secure, and in a way naturally successful, since it is enclosed by two harbors at its island, providing protection for any ship.

(337) There was once a greatly commemorated city that was notable, strong, wealthy, and beautiful, named Sybaris from the Sybaris River, and which was a famous Achaian settlement.

There were almost ten myriads in the town, and it was furnished with great abundance. But they were lifted up beyond human suitability, and the famous city was destroyed and became depopulated, since they had not learned to bear well an excess of good things. It is said that they no longer observed the strictures of the laws of Zaleukos, but having chosen luxury and a life of ease, they proceeded in time to arrogance and satiety. They were eager to abolish the Olympic contest and to remove the honors of Zeus on the pretext of instituting their own kind of gymnastic contest, with large prizes, at the same time as the one among the Eleians, so that everyone, drawn by the prizes, would eagerly come to them, abandoning Hellas. The Krotoniates, situated nearby, took them by force after a short time, after they had survived without difficulty for 190 plus 20 years.

(361) Immediately after Italia lies the Ionian Strait. The Iapygians live as far down as the entrance, and after them are the Oinotrioi and Brentesion, a port of the Messapians. Beyond them are the Keraunian Mountains. <To the west> of the Messapians live the Ombrikoi, who, they say, have chosen a delicate form of life, most resembling the style of the Lydians.

(369) Then there is the so-called Adrian Sea. Theopompos [F130] has recorded its position, and that it forms an isthmus with the Pontic Sea, having islands that are most similar to the Kyklades, including the so-called Apsyrtides and Elektrides, and also the Libyrnides. It is reported that in a circle around the Adriatic Gulf live a multitude of barbarians, nearly 150 myriads of them, who hold fine and fruitful land where they say that animals even bear twins. The air over them changes from that around the Pontos, despite it being nearby, for it is not snowy or excessively cold, yet it always remains damp. Its changes are sharp and disordered, especially in summer, with waterspouts and thunderbolts, and the so-called typhoons. There are about 50 Enetian cities lying on it at the recess, who, they say, actually came from the Paphlagonian land and settled around the Adria. There is also the Eridanos, which produces the best *elektron*, and which they say is tears turned to stone, something translucent trickling down from the poplar. They say that formerly the thunderbolt against Phaethon occurred here, which is why the great number of locals wear black and mourning clothing.

(398/391) After the Enetoi are the Thracians called the Istroi. There are two islands lying alongside them, which seem to produce the best tin. Above them are the Ismenoi and the Mentores. The region lying next to them is occupied by the Pelagonoi and the Libyrnoi. Attached to these is the ethnic group of the Boulinoi, and then the great peninsula of Hyllike, which is about equal to the Peloponnesos. They say that the

Hylloi occupy 15 cities within it and are Hellenes ethnically. They took Hyllos the son of Herakles as their founder, but it is recorded that they became barbarized by the neighboring peoples, as Timaios [F77] and Eratosthenes [*Geography* F146] say. There is an island near them called Issa, which has a Syrakosian settlement.

(415) The Illyrian land extends after this, and has many peoples. They say that the Illyrians are especially numerous, and that some of them possess the interior, where they live, and others have the coast within the Adria. Some are under royal control, others have monarchies, and some are autonomous. They say that they are exceedingly pious, very just, and hospitable, and they love sociability and strive after a most well-ordered life. Pharos Island lies not far from them, a Parian foundation, and also what is called Black Korkyra, which the Knidians settled. This territory has a certain lake that is quite large, which they called Lychnitis. Adjacent to it is an island where some say that Diomedes came and ended his life, and thus it is called Diomedeia. Above them are the barbarian Brygoi. On the sea is Epidamnos, a Hellenic city, which seems to have been founded by Korkyra. Above the Brygoi live the so-called Encheleioi, whom Kadmos once ruled. Adajacent to it is Apollonia, a Korkyraian and Corinthian foundation, and Hellenic Orikos, a coastal city, which the Euboians founded upon their return from Ilion, carried along by the winds.

(444) Next, the barbarian peoples of the Thesprotoi and Chaones do not have much territory. The island of Korkyra is next to Thesprotia, and after the Thesprotoi live the so-called Molottoi, whom Pyrrhos, the child of Neoptolemos, once brought down here. There is also Dodone (the oracle of Zeus), which was a Pelasgic establishment. Mixed barbarians are in the interior, who are said to live near the oracle. After the Molottoi is Ambrakia, a Corinthian settlement, founded by Gorgos, the oldest child of Kypselos. Then there is the so-called Amphilochian Argos, which seems to have been founded by Amphilochos, the son of the prophet Amphiaraos. Barbarians live above them. On the coast is the city of Anaktorion, which the Akarnanians and Corinthians founded. Akarnania is after it. Some of them, they say, were settled here by Alkmeon, and others were established by his son Akarnan. There are a number of islands near here: Leukas, one of the first, was a Corinthian foundation. Then there is that of the Kephallenians, and nearby Ithake, with Zakynthos lying near to the Peloponnesos. Then, extending toward the Acheloos, are the so-called Echinades.

(470) Next we will go back through Hellas. We will show in summary form all the places around it in ethnic fashion, and

according to Ephorus [F144]. After the Akarnanians is Aitolia, which took its settlers from Elis. The Kouretes fomerly lived there, but Aitolos, arriving from Elis, named it Aitolia and expelled them. The city of Naupaktos lies near Rhion, and was founded by Dorians with Temenos. After the Aitolians are the Lokrians who live alongside them, those who are called the Ozolians, having been settled from the Lokrians who are turned toward Euboia. Delphi is attached to them, and the most trust-worthy Pythian oracle. Then there are the Phokians, whom Phokos seems to have unified, having come earlier down to it with Corinthians, and who was descended from Ornytos the son of Sisyphos.

(488) Boiotia lies next to this, an exceedingly large territory that is fortunate in its position, for, as the story is, it alone has the use of three seas. It has harbors that look most favorably toward the midday, toward the Adria and the Sikelian emporion, and to Cyprus and the sail to Egypt and the islands. These places are around Aulis, and among them lies the Tanagraian city, with Thespiai above it in the interior. The third is outside the course of the Euripos and leads to the Macedonians and Thessalians, with the coastal city of Anthedon next to it. Thebai is the largest in Boiotia.

(502) Megara adjoins next, a Dorian city, for all the Dorians together–mostly Corinthians and Messenians–made it a city. They say that Megareus, the son of Onchestos, became its ruler and put his name on it. The Megaris borders Boiotia. The Cor-inthian Gulf is next, and the Kenchrean, which draw the isthmus together into a narrows, with the mainland on either side. Then the Peloponnesos lies next, which has deep gulfs and many capes, including Malea, and the largest, called Tainaron, where there is a sanctuary of Poseidon, most suitable for the god and established by the Lakonians. The portions of the Peloponnesos toward the *boreas* are held by the Sikyonians and those who once settled the distinguished city of Corinth, as well as the Achaians. The boundaries toward the *hespera* and *zephyros* are Eleian and Messenian. Those toward midday and the southern *klima* are Lakonian and Argive, and that toward the sunrise is held by the cities of Akte. Phleiasia is in the interior, as well as the Arkadian peoples, the greatest of them. They say that the Arkadians are autochthonous, and that later Aletes settled the Corinthians, with Phalkes Sikyon, Tisamenos Achaia, Oxylos became the leader of Elis, Kresphontes of Messenia, Eurysthenes and Prokles of Lakedaimon, Kissos of Argos (along with Teme-nos), and Agaios—the story goes—of those around Akte, along with Deiphontes <the relative of> Temenos.

(535) The island of Crete lies opposite the Peloponnesos. It is large in size and especially fortunate, extending from Maleia—

the Lakonikan cape—lengthwise into the open sea as far as Dorian Rhodes. It was inhabited almost from the beginning by large crowds and cities. Their oldest settlers are called the Eteocretans. They say that the Cretans were the first to rule the Hellenic sea and to take over the island cities. Ephorus [F145] says that they populated some of them, and adds that the island was named after a certain Kres, who was autochthonous and became king. It is a day's sail from Lakonike.

(550) Lying in the Cretan Strait is Astypalaia, a Megarian settlement that is an island in the open sea. Near to Lakonike is Kythera, and after these, near Epidauros, is what was formerly called Oinone, but later Aiakos took it and called it Aigina after Aigine the Asopid. Near to it is Salamis, about which the story is that Telamon the son of Aiakos once ruled it.

(559) Next there is Athens. They say that it first acquired Pelasgians as settlers, the ones, as it is reported, who were called Kranaoi. After them there were the Kekropidai, when Kekrops held power, and, in later times, Erechtheus ruled the city and it acquired its name from Athene. Herodotus [8.44] records these things in his history.

(566) Going around Sounion, after Attike there lies Euboia Island, which was formerly called Makris [Long], as they say, because of its nature. In time it eventually took the name of Euboia from the so-called Asopid. They say that formerly the first settlers were mixed Leleges, who lived together there, but Pandoros the son of Erechtheus crossed from Attike and founded Chalkis, the greatest city on it. Aiklos, of Athenian ethnicity, founded Eretria, and Kothos similarly Kerinthos on the sea, the Dryopes what is named Karystos, and Hestiaia was a foundation of the Perrhaibians. Lying nearby are the islets of Skyros, Peparathos, and Skiathos. The Cretans who once crossed over from Knossos with Staphylos populated Peparethos and Ikos, an island lying near it, and Skyros and Skiathos were populated by Pelasgiotes who crossed over from Thraikia, as it is reported. When all of these became deserted the Chalkideans again populated them together.

(587) The Lokrians live opposite Euboia. The first to rule them, as they say, was Amphiktyon the son of Deukalion, and the next by blood was Itonos, and then Physkos, who produced Lokros, who named the Leleges the Lokrians after himself. After them there are the small Dorian cities of Erineos, Boion, and Kytinion, which are all exceedingly ancient, and then Pindos. These were settled by Doros, the son of Hellen. All the Dorians are emigrants from them. Next, near these, is the city of Herakleia, which the Lakones formerly founded, sending a myriad of

settlers into Trachis. Next from there is coastal Pylaia, where the Amphiktyony assembles. The Malaic Gulf lies in the recess, where the city of Echinos is—founded by Echion the Sown—as well as other Malian cities. Then there are the coastal Phthiotic Achaians. The Magnetes live around Pelion. Above them is a territory that is abundant in pasturelands and which has masterful and fruitful plains, as well as Larisa, a most fortunate city, and many others. The course of the Peneios, a great river, flows through the narrows of Tempe and the deep-water lake near Pelion called Boibeis. Athamania borders on Thessaly and the adjacent Dolopian and Perrhaibian peoples, as well as the Ainianians, who seem to have originated from the Haimonians, Lapiths, and Myrmidons.

(618) Above Tempe is the Macedonian territory, which lies next to Olympos. They say that earthborn Makedon ruled it. The Lynkestoi and also the Pelagonian peoples lie there, alongside the Axios, and the Botteatai are near the Strymon. In the interior there are many cities, with Pella and Beroia the most distinguished. On the coast are Thettalonike and Pydna. Going around the cape called Aineia, there is the former Corinthian foundation of Potidaia, a Dorian city afterward named Kassandreia. In the interior is what is called Antigoneia, and later there was the city of Olynthos, which Philip [II] the Macedonian razed after conquering it by the spear. After the Olynthia are Arethousa and Pallene, which is located on an isthmus. It is believed that this was formerly called Phlegra, according to the story, which was settled by the Giants who fought the gods. Later the Pelleneai, landing here from Achaia, named it after themselves. Next is the gulf called the Toronic, where the city of Mekyberna was formerly located, and then there is Torone, homonymous with these places. Then in the open sea there is Lemnos (the nurse of Hephaistos), settled first by Thoas the son of Dionysos. Afterward it had Attic immigrants. Upon sailing past Athos there is Akanthos, a coastal city and an Andrian settlement. Next to it a cut ditch of seven stadia is pointed out, and it is said that Xerxes cut it. Then there is Amphipolis. The Strymon, a great river, flows beside it, carried as far as the sea by the so-called Dances of the Nereids. From there, in the interior, is the homeland of Antiphanes, called Berga, who wrote an unbelievable and laughable mythic history. After Amphipolis is the city that was formerly Oisyme and which was Thasian and later Macedonian, called after the Makessa Emathia. Next is Neapolis and the island of Thasos, which, as the story goes, was first founded by barbarians and then Phoenicians who crossed from Asia with Kadmos and Thasos, from whom it took the name Thasos, which it has today, from Thasos.

(664) The territory beyond is held by Thracians, who extend as far as the Pontic Istros. Those who lie on the coast have a city, Abdera, named after Abderos, who once founded it and who, it seems, was later destroyed by the horses of Diomedes, the killer of strangers. The city was repopulated by the Teians, who were fleeing because of the Persian matter. Beyond it there is situated a river called the Nestos. In the parts toward the sunrise is an elongated lake, Bistonis, taking its name from the Bistonian Thracians. Then there is Maroneia. It is recorded that the Kikones—those in Ismaros—formerly settled there, and that it was a foundation of the Chians.

(679) Opposite is Samothraike, a Trojan island that has a mixed population of settlers. Formerly, some say, the Trojans were there. Elektra, called the daughter of Atlas, gave birth to Dardanos and Iasion. Regarding them, Iasion committed a sacrilege against an image of Demeter and was struck by a divine thunderbolt and died. Dardanos had <previously> left these places and...founded at the foot of <Mt.> Ida a city called Dardania, after himself. The Samothraikians, who were ethnically Trojan but were called Thracian after the place, remained at the place out of piety. Once, during a famine, they were provisioned by the Samians, and at that time received some immigrants from Samos and took them in.

(696) After Maroneia lies the city of Ainos, which has Aiolian settlers from Mytilene. The peninsula of Thraikia lies next, on which the first city is Kardia, originally founded by Milesians and Klazomenians, and then later by Athenians, when Miltiades took power over the Chersonesians. Then there is Lysimacheia, the one that Lysimachos founded and named after himself. Next is Milesian Limnai, and then the Aiolian city of Alopekonnesos, and after it Elaious, an Attic settlement, which Phorboon [?] seems to have founded. Then there are Sestos and Madytos, lying on the narrows and founded by the Lesbians. Then there is the city of Krithote, and Paktye, both of which they say that Miltiades founded.

(713) After the Chersonesos, Thraike extends to the Propontis, and there is a Samian settlement, Perinthos. Then there is Selymbria, which the Megarians founded earlier than Byzantion, and then prosperous Byzantion, which is Megarian. After this there is the Pontos, whose situation seems to have been studied most carefully by the writer from Kallatis, Demetrios [T4]. We will go through its places by parts. Near the mouth of the Pontos is the Byzantine territory, called Philia. There there is a coast called Salmydessos, which is shallow for 700 stadia, and whose extent is exceedingly difficult to land on. It is completely without harbors and a place totally hostile to ships. Cape

Thynias, with a good harbor, is attached to it, and it is the farthest place in Astic Thraike, after which there is the neighboring city of Apollonia. The Milesians founded this city, when they came to this place about fifty years before the reign of Kyros, for they sent out many settlements from Ionia to the Pontos, which had formerly been called Axenos [Inhospitable] because of barbarian attacks, but when they arrived they made it the Euxeinos [Hospitable]. At the foot of the mountain called Haimos is the city named Mesembria, which borders the Thracian and Getian land. It was settled by Kalchedonians and Megarians, when Dareios [I] made his expedition against the Skythians. There is a great mountain above it, Haimos, similar to the Kilikian Tauros in size and the lengthwise extension of its localities. From the Krobyzoi and Pontic boundaries it goes as far as the region of the Adriatic.

The extant text terminates at this point, after 747 lines, with the description of Mt. Haimos. The *Periodos* is the last item in the manuscript, which is generally of poor quality, and its final pages are missing, which would also have included the name of the author. Yet in 1628 Lucas Holsten recognized that another text, the *Periplous of the Euxine Sea*, preserved in its entirety in Codex Vatopedinus 655 from Mt. Athos, contained verbatim the last 26 lines of the extant *King Nikomedes Periodos*, beginning with the citation of the Byzantine territory near the mouth of the Pontos at line 722.[23] The *Periplous of the Euxine Sea* is a compilation from several sources, including the extant *Periplous of the Euxine Sea* of Arrian, as well as the *Periodos*, and describes the coast of the Euxine (Black) Sea. Holsten realized that direct quotations from the *Periodos* were embedded in the text, often identifiable by the use of iambic meter.

Various attempts to reconstruct the otherwise lost portion of the *Periodos* have been attempted since the time of Holsten, especially by Aubrey Diller in 1951, who was able to provide an additional 278 lines of the text, although not without gaps and problems. These created a narrative that proceeded in a clockwise fashion around the Black Sea, with some divergences away from the coast. The account ends at the northwest corner of Asia Minor on the eastern side of the Thracian Bosporos, having made a complete circuit of the sea. This was not the end of the *Periodos*, since its broader scope was stated at the beginning (lines 73–97). It would have included the western coast of Asia Minor (and thus Bithynia, the homeland of King Nikomedes), and also the coasts of Syria, the Levant, and North Africa, ending where it began, at the Atlantic mouth of the Mediterranean.

Diller's reconstruction is presented below. Toponyms and ethnyms in brackets appear in the *Periplous of the Euxine Sea* but are not in metrical form, so they are not direct quotations from the Nikomedes *Periodos*.

(748) The Milesians were founding [the city of Odessos] when Astyages ruled Media, which has Thracian Krobyzoi in a circle around it. [Dionysopolis] was formerly called Krounoi because of the outflow of nearby water, and later, when the Dionysiac statue fell from the sea onto the place, it was renamed Dionysopolis. On its boundaries are the Krobyzoi and the Skythians; it has mixed Hellenic settlers. Some say that [the city of Bizone] is barbarian, and others that it was established as a Mesembrian settlement. [The city of Kallatis] was a Herakleotic settlement, and was founded according to an oracle when Amyntas took over the rule of the Macedonians. [The city of Tomeoi] was established as a Milesian settlement, with Skythians settled around it. [The city of Istros], which took its name from the Istros River, was being founded by the Milesians when the expedition of the barbarian Skythians crossed into Asia, chasing the Kimmerians from the Bosporos.

(771) [The Istros River] comes down from the western regions and makes its outlet in five mouths, but it also splits in two into the Adria. It is in fact known as far as Keltike, and endures all the time, even in summer. In winter it is increased, filled by the rains that occur from the snow, as they say, taking its inflow from the melting frost, yet in summer it puts forth a stream that is exactly equal. It has islands lying in it, which are numerous and of great size, as the story goes. The one lying between the sea and the mouths is no smaller than Rhodes. It is called Peuke because of its great number of pines. There is also the island of Achilles after it, lying in the open sea. It also has a tame mass of birds, and is seen as a sacred place to those arriving. One cannot see the mainland from there, since it is 400 stadia from land, as Demetrios [F2] writes... . Thraikes and immigrant Bastarnai... . [The Tyras River] is deep and abundant in pasture for grazing, and has places for fish markets and a safe upstream sail for merchant ships. A homonymous city lies on the river, Tyras, which was a Milesian settlement.

(801) [The Borysthenes River] is the most useful of all, with many large fish, crops that grow along it, and pastureland for herds. They say that its stream is navigable for 40 days, although its upper portions are not navigable or passable because it is blocked by snow and ice. At the confluence of the Hypanis and Borysthenes Rivers a city was founded, formerly call Olbia but later called Borysthenes again by the Hellenes. The Milesians founded it during the Median hegemony. It is a sail of 240 stadia upstream from the sea, by means of the Borysthenes River. [The Racecourse of Achilles], which is a beach, is exceedingly long and narrow. Some say that Iphigeneia came to [the Taurian people] after her abduction at Aulis. The

Taurians are numerous and crowded together, and strive after a pastoral mountain life, but they are barbarian in their savageness and also murderous, placating the divinities for their sacrileges. What is called the Tauric Peninsula is attached to these, having a Hellenic city that was settled by the Herakleots and Delians, since an oracle came to the Herakleots—who lived in Asia within the Kyaneans—that they were to settle the peninsula along with the Delians.

(832) It is said that [in the city of Theodosia] exiles from the Bosporos were settled. In the sea opposite [the city of Kimmerikon] are two rocky islets that are not very large and a little distance from the mainland. The farthest [city] is [Pantikapaion], which is designated the royal residence of the Bosporos. Above them there is barbarian Skythike, bordering on the uninhabitable land and completely unknown to the Hellenes.

(842) The first people along the Istros are the Karpides, as Ephorus [F158] says, and then the Aroteres are farther away, with the Neuroi as far as the territory that is deserted because of frost. Toward the sunrise and across the Borysthenes are the Skythians living in what is called Hybla, with the Georgoi above them. Then it is deserted again over a wide extent, and above are the Anthrophagian Skythian people, and then it is deserted again. After crossing the Pantikapes, there are the Limnaian people and several others who are not distinguished by name but are called pastoralists. They are exceedingly pious, and none of them would commit injustice on a living person. It is said that they carry their houses and feed upon the milk of the Skythian mare milkers. In their lifestyle they consider possessions to be common to everyone, as well as all social intercourse. He says that wise Anacharsis was born from the most pious pastoralists.

(860) ...some came to Asia and settled there, whom they call the Sakai. He says that the most conspicuous are the Sauromatai and Gelones, and third are the people surnamed the Agathyrsoi. Taking its name from the Maiotai, the Maiotis Lake lies next, into which the Tanais...which takes its flow from the Araxis River and is mixed with it, as Hekataios of Teos says, or, as Ephorus [F159] recorded, from a certain lake whose limit is unperceived. It empties into the so-called Maiotis in a double-mouthed stream, and then into the Kimmerian Bosporos.

(875) The Tanais, which is the boundary of Asia, cuts each continent apart. It is first held by the Sarmatai, who extend for 2000 stadia. Then there are the Maiotai people called the Iazamatai, as Demetrios [F1] has said, but according to Ephorus [F160] they are the Sauromatian people. They say that Amazons mingled with the

Sauromatai, when at one time they came from the battle that occurred around the Thermodon, and because of this they were surnamed
the Gynaikokratoumenoi. Then there is Hermonassa, and Phanagoreia, which, they say, was once settled by the Teians, and Sindikos
Harbor, whose founders were Hellenes who came from nearby places.
These cities are enclosed by the island on which they lie, along the
Maiotis as far as the Bosporos, which consists of an extensive plain,
and which is impassible due to marshes and small rivers, as well as
shallows in the farthest part, which are formed by the sea and the
lake.

(895) As one sails out of the mouth, there is the city of Kimmeris,
called after the barbarian Kimmerians, founded by the *tyrannoi* of
the Bosporos. There is also Kepos, settled by the Milesians. [The
Sindian people] are among the Maiotai, who are barbarian but
peaceful in their customs. [The Kerketai] are a just and reasonable
people, and are excellent sailors. Those called the Achaioi hold the
land beyond, bordering on them, who, they say, are ethnically Hellene and are called the barbarized Achaioi. They say that at one time
the Orchomenian population under Ialmenos and the Minyans, sailing with all their army from Ilion, came unwillingly into the Pontic
and barbarian territory because of the wind. Thus, they say, having
been driven into exile, they were lawless and especially hostile in their
customs toward the Hellenes. Many of the Achaioi are opposed to
the Kerketai.

(914) [The Heniochian people] hate foreigners. Some say that they
were named from the charioteers of Polydeukes and Kastor, Amphitos and Telchis. They seem to have arrived on the expedition with
Jason, but settled around these places after being left behind,
according to the story. Above and inland of the Heniochoi lies the
sea called the Kaspia, which has horse-eating barbarians living
around it. The boundary of the Medes is nearby. The stream [of the
Phasis River] is carried from Armenia, near to which Iberians live
who were removed from Iberia to Armenia. Upon going upstream,
on the left of the Phasis is the Hellenic city of the Milesians called
Phasis, lying alongside it. The story is that 60 peoples come down
into it, who use different languages. They say that barbarians from
Indike and the Baktrianian territory come together among them. After
these is barbarian Koraxike, and then those called the Kolike, the Melanchlainoi, and the Kolchoi people. [The Makrones people]... .[The
Mosynoikoi people] have savage customs and are most barbaric in their
actions. They say that they all live in exceedingly high wooden towers,
but everything they do is visible to all, and their king, who is bound and
shut up in a tower, watches over them carefully and has the highest roof.

Those guarding him take care that everything he commands is lawful, and if he transgresses, he is given the greatest chastisement, they say, and is not provided with food.

(950) [The city of Pharnakia] was founded by...and is near a deserted place [?], opposite which is what is called the Island of Ares. [The Tibarenian people] are neighbors and are eager to play and laugh in every way, since they have judged this to be the greatest happiness. [The city of Amisos] lies in the land of the Leukosyroi and is a settlement of <the Milesians> and Phokaians; since it was founded four years earlier than Herakleia, it took an Ionian foundation. Next to this city is what is almost the narrowest part of Asia, extending to the Issic Gulf and Alexandroupolis, founded by the Macedonian. There is a road of seven days total to Kilikia, for it is said that this is the most isthmus-like portion of Asia, where it comes together in a recess. Yet Herodotus did not seem to know about it, saying that there is a straight route of five days beginning from Kilikia and going as far as the city of Sinope, as he records in his writings [2.34]. The peninsula mixes together almost all the best lands in Asia, and has 15 peoples, of which three are Hellenic (Aiolic, Ionic, and Doric). Otherwise the remaining mixed populations are barbarians. Living here are Kilikians, Lykians, and also Karians, the Mariandynians along the sea, Paphlagonians, and Pamphylians. There are also Chalybians in the interior, and the Kappadokians near them, those holding Pisidia, the Lydians, and also the Mysians and Phrygians.

(981) [The Halys River] is 300 stadia from Amisos, running between the Syrians and the Paphlagonians, emptying into the Pontos. [The city of Sinope] was named after one of the Amazons who once lived nearby and was related to the Syrians. After that, as they say, there were the Hellenes who crossed over to go against the Amazons: Autolykos and Phlogios (with Deileon), who were Thessalian. Then there was Abron, ethnically Milesian, who seems to have been killed by the Kimmerians. After the Kimmerians there were Kretines the Koan, and those who were exiles from the Milesians. They were founding it when the Kimmerian army came down into Asia.

(998) [The promontory of Karambis] is a high mountain cut precipitiously into the sea. Krioumetopon is a sail of a night and a day from Karambis. They say that Phineas, the son of Phoinix the Tyrian, ruled the places around [the city of Amastris], and in later times an expedition of Milesians came from Ionia and founded these cities, places which Amastris later brought together and founded the city of the same name, Amastris. She was reported to be the daughter of Oxathres the Persian, as the story goes, and the wife of Dionysios the *tyrannos* of Herakleia.

(1012) [The Parthenios River] is navigable and descends with an exceedingly quiet current...the story is that there are very famous baths of Artemis in it. [The city of Herakleia] is a Boiotian and Megarian foundation, founded within the Kyaneai by those setting forth from Hellas at the time that Cyrus took over the Medes. [The Hypios River] has an inland city on it called Prousias. [The Sangarios River], flowing from above the Thynoi and the Phrygian land, empties through Thynis. [The island of Apollonia] has a city on it called Thynias, a foundation of Herakleia.

Commentary

Lines 1–15

In an introduction of 15 lines, the author made his appeal to King Nikomedes and immediately stressed the legitimacy of his use of comic meter. He may have had the prologues of comedy in mind, given the word *komoidia* in the first line, as well as *psychagogein* in the fourth, the latter originally referring to the leading of souls from the underworld but eventually having the more metaphorical meaning of "pleasurable."[24]

The implication is that the author was a member of the Bithynian royal court, or perhaps an aspirant, which was located at Nikomedeia (modern İzmit in northwest Turkey). The city had been founded in the 260s BC by Nikomedes I, using the standard Hellenistic process of bringing various rural and village populations together into a new settlement. Only scattered remains are visible, mostly of the Roman period.[25]

The introduction is an appeal to the king for an interview, which presumably would include a presentation of the treatise, stressing his patronage of scholarship, about which nothing else is known. Yet the author (perhaps considering the interview as much as the text) emphasized that he would speak in the "Lakonikan manner." Since at least the fifth century BC, the traits of Spartan speech were proverbial, giving rise to the word "laconic," one of their many customs that other Greeks found peculiar. The author thus demonstrated that he did not intend to impose seriously on the king's time, either in person or with his treatise.

Lines 16–44

In order to strengthen his credentials, the author associated himself with Apollodoros of Athens and his teachers. The Pergamene kingdom ended in 133 BC with the death of Attalos III and his legacy to Rome, and the implication is that this passage was written some years later, probably after the accession of Nikomedes III in the early 120s BC.

Apollodoros was not mentioned by name in the extant *Periodos*, following a tradition of speaking of one's illustrious predecessors allusively. Empedokles used the same technique in refering to Pythagoras, and perhaps the most famous example is Lucretius' diffidence in naming Epikouros.[26] Nevertheless, in the *Periodos* there is no doubt that Apollodoros was the personality meant. He was one of the leading scholars of the second century BC, and was still alive or recently dead when the *Periodos* was produced. He recorded the campaign against the Allobroges by Quintus Fabius Maximus in 122/121 BC, and the Athenian archonship of Eumachos of 120/119, the latest events preserved in the fragments of his works.[27] If he were still living when the *Periodos* was written, this may have been one of the reasons he was not mentioned by name, in order to avoid the impression that he, rather than the king, was the true dedicatee.

Apollodoros was from Athens, but spent a number of years in Alexandria. He left there after the accession of Ptolemy VIII in 145 BC and his repression of scholarly activities, and went to Pergamon, dedicating his *Chronika* to Attalos II, who reigned until 138 BC. A typical Hellenistic polymath, he wrote on Homeric studies (with a topographical slant), especially an account of the *Catalogue of Ships* that was to be heavily used by Strabo in Books 8–10 of his *Geography*. But Apollodoros is best remembered for his four-book *Chronika*, a verse account of events from the Trojan War to his own day, based upon but superseding the similar work by Eratosthenes.[28]

The author of the *Periodos* also drew attention to Apollodoros' education in Athens, thus making a subtle reference to the intellectual tradition of that city. He further strengthened his position as a legitimate member of the scholarly elite of the second century BC by naming two of Apollodoros' teachers, Diogenes and Aristarchos. The former was from Babylon and became the head of the Stoic school, and also brought Stoicism to Rome in the mid-second century BC.[29] Aristarchos of Samothrace, active in the first half of the second century BC, was head of the Library in Alexandria, but, like Apollodoros, left at the time of the expulsions by Ptolemy VIII. He was a grammarian and literary commentator, and noted for his high level of scholarship.[30] By aligning himself with these three recent scholars, the author of the *Periodos* made it clear to King Nikomedes that he himself was worthy of respect.

Moreover, there are two casual words in this portion of the treatise that illuminate the author's intentions. *Philologos* (line 19) had existed since at least the fourth century BC, and meant a highly educated person, in time particularly applied to Eratosthenes, who allegedly was the first to use it as a surname.[31] In the *Periodos* this was applied to Apollodoros, thus making a subtle connection to Eratosthenes and the origins of geographical scholarship.

After associating himself with the best in contemporary and previous scholarship, geographical and otherwise, the author used a more contemporary word, *chronographia* (line 23) in order to place himself within the scholarly writing of his own era. This was a neologism, first cited in extant literature by the author's older contemporary Polybios and meaning an informal but chronologically arranged account of events.[32] It is possible that Apollodoros also used the term, but it does not appear in the numerous fragments of his works. Thus by astute choice of diction, the author of the *Periodos* located himself not only within the best of Hellenistic scholarly traditions but its later innovations. Yet by not mentioning Apollodoros by name, he retained a certain elusiveness that demonstrated his independent profile.

He also oriented himself toward the most distinguished and cultural state of Asia Minor, Pergamon (lines 16, 45–9). Although Pergamon had passed from the scene as an independent entity in 133 BC, and was in the process of evolving into a Roman province when the *Periodos* was written, it retained the intellectual predominance for which it had become famous, including the second-best library in the Hellenistic world, established by Eumenes II in the first half of the second century BC.[33] Pergamon was outstanding culturally in other ways, with its patronage of art and architecture, and Pergamene monuments were scattered across the Greek world, most notably the two royal stoas still visible in Athens, one of which was the work of Attalos II (ruled 159–138 BC), the patron of Apollodoros mentioned by name in the *Periodos* (line 48). It was no accident that scholars expelled from Alexandria by Ptolemy VIII ended up at Pergamon.

While at Pergamon, Apollodoros completed at least the major part of his *Chronika* and dedicated it to Attalos II. The *Periodos* provides a catalogue of the topics that it included, as well as its chronological extent, from the Trojan War to "the present time," which would be before 138 BC, the death of Attalos II, at least for the portions dedicated to him. A span of 1040 years back to the Trojan war would seem to conform with Eratosthenes' date equivalent to 1184 BC.[34] The catalogue in the *Periodos* lists 14 topics included in the treatise: as the author put it, "a summary of everything." It is a particularly astute outline of the basis of Greek ethnography and history as it was understood in the Hellenistic period, casting a broad net from political and military events to biography, cultural institutions, and the movements of non-Greek populations.

The author of the *Periodos* also felt the need to remind his readers once again about the virtues of the metrical form, although in this case applying it to Apollodoros rather than himself. He used the unusual analogy of the difference between carrying loose pieces of wood and having them bound together. Since the beginnings of geographical scholarship it had been common to explain abstruse concepts by household terminology and analogies: Eratosthenes used the spindle whorl and the chlamys to provide templates for sections of the inhabited world.[35] Apollodoros' analogy is literary, not scientific, but nevertheless a feature of the same process: making a technical discussion more viable to non-specialists. In fact, the wood carrier anecdote in the *Periodos* may be a witty reference to a situation from comedy.

Lines 45–64

Apollodoros submitted his *Chronika* to Attalos II Philadelphos of Pergamon, although since the king died in 138 BC and there appear to be items in the treatise as late as 120 BC, he continued to work on it for another two decades. Presumably the "Attalos" named is the same king, but there is a remote possibility that it was his nephew and successor Attalos III (ruled 138–133 BC), the last king of Pergamon. Yet the author of the *Periodos* showed no knowledge of Apollodoros' work on the *Chronika* after the death of Attalos II.

With the collapse of the Pergamene monarchy, the status of Bithynia was raised somewhat, since it was the only remaining kingdom on the Aegean coast of Asia: hence the words of praise for King Nikomedes. The only rival emerging was the Pontic kingdom of Mithridates VI, but he had just come to the throne when the *Periodos* was written and was an adolescent contending with internal instability (his father had been assassinated). Thus until Mithridates consolidated his power, Bithynia was the most powerful state in western Asia. The *Periodos* reflects this era of Bithynian prominence under Nikomedes III, and comparing the king to Attalos II, arguably the greatest monarch of the second century BC, only reinforces this. The author, who noted that he would report the distinction of Nikomedes to others, thus became a spokesman for the regime.

To strengthen this, the author invoked Apollo of Didyma, the great oracle just south of Miletos, whose visible remains are one of the spectacular archaeological sites of the region. Although there is evidence for a relationship between the Bithynian royal family and Didyma, there is nothing else about any specific assistance given to the father of the king, but obviously the author felt that there was something he could draw upon to flatter the king and enhance his own status.[36]

Lines 65–108

The author continued his address to the king by summarizing the topics in his treatise. The account begins with a specific notice of the major subjects covered: the first two, settlements and the foundation of cities, are a direct reference to Timaios of Tauromenion, one of the author's sources (line 126).[37] Mention of "places that are navigable and passable" demonstrates indebtedness to the *periplous* format. The account is oriented toward chorography (with some geographical material), in other words, a description of places and their characteristics. The author did not use the word *chorographia*, perhaps because the concept was a new one in his day, probably the invention of Polybios.[38] Yet he emphasized that he would draw his material together from "scattered" (*sporaden*) sources, an agricultural image of sowing and reaping, demonstrating not only how he was drawing everything together but also producing a finished product.[39]

The list is a standard catalogue of chorographical data, but there is the peculiarity that the author stated there were only two continents, Europe and Asia (lines 76, 87–8). Continental theory had originated in the fifth century BC, perhaps with Hekataios of Miletos. In fact, there were originally two— Europe and Asia—but the third, Libya (Africa) was assumed by late in the century.[40] The reversion to two by the author of the *Periodos* is an anomaly, and may suggest that he did not consider Libya worthy of comparison with the other two: it is perhaps of interest that any Libyan portion of the treatise, whatever it might have been, has vanished.

Without stating it explicitly, the author hinted at the utility to royalty of chorographic and geographical information, noting that his treatise would provide the king with an account of the entire inhabited world. Hellenistic royalty and geographical knowledge had been intertwined since the days of Alexander the Great, whose expedition to India substantially expanded the known limits the inhabited world. Given the fluid boundaries and expansionist policies of the Hellenistic states, it was obvious that royalty and people of power needed to be astute geographically, a point of view to be expressed most vigorously by Strabo, who argued at the very opening of his *Geography* that geographical knowledge was necessary "for political activities and those of commanders."[41] In expressing such views to the king, the author of the *Periodos* used the technical word *oikoumene* ("inhabited region"), a term developed by Aristotle that became standard ever thereafter among geographers to contrast the small portion of the earth that was inhabited with the vast region that was deserted.[42] Thus the author demonstrated that he expected the king to be aware of the vocabulary of geographical thought. Yet, fearing that he might impose on the king's time, he again stressed the brevity of his treatise (line 91).

As a final way of emphasizing the cultural importance of the *Periodos*, the author invoked the name of the first human traveller and explorer, Odysseus, paraphrasing the opening of the *Odyssey* with "towns and customs" (line 102). The introduction of the hero into the narrative is somewhat unexpected, yet it sets the king in opposition to him: the king will not have to wander through the world, nor (given the brevity of the treatise) have to listen to an endless account of the wanderings of someone else, a wry allusion to the nature of much of the *Odyssey*.[43] Returning to flattery of the king (now linked with the heroic age of Greece), the author's final address to him noted that dissemination of the treatise will also mean the spread of his name.

Lines 109–36

The next section of the *Periodos* lists the author's major sources. Unfortunately this is one of the most corrupt portions of the manuscript, with two lines totally illegible and five others only partially readable (lines 119–25), so the catalogue of authors is woefully incomplete. What survives is a list of nine authors from the fifth century BC to essentially the author's own time. First, expectedly, is Eratosthenes, the inventor of the discipline of geography and much of its technical terminology. His three-book *Geographia* was published in the second half of the third century BC.[44] By placing Eratosthenes at the head of his list, with the name at the beginning of its line in the text, the author left no doubt that his own work was in the mainstream of geographical scholarship.

The following line (115) begins with the name of another important source, Ephorus of Kyme, the author mentioned most frequently in the surviving text. Writing a century before Eratosthenes, he created the first universal history and also innovatively included a section on geography within his work, thereby giving attention to many of the ethnographical topics that were of interest to the author of the *Periodos*.[45] It is clear that Eratosthenes and Ephorus were an important basis of the research for the *Periodos*, especially the latter.

The next few authors are somewhat more vague. Dionysios of Chalkis is an obscure Hellenistic author, whose only known work was titled *Foundations*.[46] Yet given the extensive information about city foundations in the *Periodos* he was presumably an important source. Demetrios of Kallatis (a town on the west coast of the Black Sea), from the early second century BC, wrote 20 books on Asia and Europe, perhaps with an emphasis on the northern portion of the inhabited world.[47] Kleon of Sicily may be the scholar from Syracuse mentioned by several sources, perhaps of the early Hellenistic period, but he is hardly known.[48] Timosthenes of Rhodes was the naval chief of staff for Ptolemy II, in the first half of the third century BC; his *On Harbors*, a nautical guide for Ptolemaic seamen, survives in about 40 fragments and would have been an important resource for seaports.[49]

After the name of Timosthenes is the illegible portion of the manuscript. Since there are two sources in each preserved line, as many as 10 could be missing. This probably included the three cited elsewhere in the text, Antiphanes of Berga, Hekataios of Teos, and Theopompos of Chios (lines 370, 653, 870), as well as several others.

When the coherent text resumes, the next name is Kallisthenes, the historian from Olynthos who was a student of Aristotle's and a companion of Alexander the Great. He wrote a history of events in the fourth century BC and an account of the eastern expedition, but since he is not mentioned elsewhere in the extant *Periodos* there is no indication of how the author might have used his material.[50]

The following line is illegible, and the next visible name is Timaios of Tauromenion, active in the late fourth and early third centuries BC, who wrote extensively about southern Italy and Sicily, and may have been the author's main source for those regions.[51] As a fitting close to the list, the author named Herodotus, thus bracketing Eratosthenes at the beginning. It should be noted, however, that several of the names do not appear in the extant text (Dionysios, Kleon, and Timosthenes, in addition to Kallisthenes), so there is no certainty how they fit into the author's plan of research.

The author then added a personal comment, noting that a significant part of the work was based on autopsy, with a special emphasis on Italy, Sicily, and the regions around them. If "toward the west" is to be taken literally, his travels included practically all of the Mediterranean world. There is only one city mentioned, Karchedon (Carthage), leading to the possibility that he had a special relationship with it.

Lines 137–69

At line 137 the descriptive part of the *Periodos* begins in earnest, at the Pillars of Herakles, long a tradition for the starting point of geographical narratives, perhaps so established by Hekataios of Miletos at the beginning of the fifth century BC.[52] The figure of 120 stadia is the first of several distance measurements in the text; stadia are the only units used by the author. Although the length of the stadion is always uncertain,[53] the amount is perhaps 24 km. This is approximately the width of the western end of the strait separating Europe and Africa (the modern Straits of Gibraltar).

The location of the Pillars of Herakles may seem obvious, especially today, at the conspicuous twin mountains called Gibraltar and Jebel Tariq (ancient Kalpe and Abilyx), yet such was not the case in antiquity. In fact there were numerous suggestions in addition to these two mountains. There was also a mythological explanation, connected with the travels of Herakles, and an ethnographic one, that the Pillars were constructions in the Temple of Herakles (Melqart) at Gadeira.[54] The author of the *Periodos* offered another location: islets near the two mountains (30 stadia would be about 6 km.), the conclusion of Euktemon of Athens, who wrote a geographical work on the western Mediterranean.[55] At least one of the islands in question was called Hera's Island, which had a temple of the goddess, perhaps modern Palomas Island near Tarifa, or Perejil Island, to its southeast just off the opposite African coast.[56]

The following mention of Mainake is unclear. Normally it is located about 30 km. east of modern Málaga at Torre del Mar, which is hardly close to the islands around the Pillars (over 100 km. away).[57] It is odd that the author should single out this place alone of the several towns on the southern Iberian coast, especially since the following passage in the *Periodos* is about localities west of the Pillars, not east. It may be that he had access to a Massalian report about their outposts, which noted that Mainake was the farthest of their cities, in other words, the westernmost. But its location remains uncertain.[58]

Using a *periplous*, the author followed an itinerary that headed west along the Iberian coast from the Pillars. The "cape opposite to the setting sun" is probably the Promontory of Hera (or Juno, modern Cape Trafalgar), which points slightly east of south and is the most prominent headland between the Pillars and the region of Gadeira. The island of Erytheia was located in a variety of places, including well out into the External Ocean, but was more commonly placed around Gadeira.[59] It was associated with the story of the cattle of Geryon, the tenth labor of Herakles; there may have been a local bull cult, with the place having an unusual density of bovines. The comparison with those from Egypt and Thesprotia (in northwest Greece) is probably purely generic rather than referring to any specific taxonomy.

"Western Aithiopians" was a generic name for peoples in the remote southwestern portions of the inhabited world. Generally these Aithiopians were confined to Africa, but localization was vague, and Ephorus seemed to place them all along the Atlantic coast.[60] At line 159 at least one word is illegible, probably a descriptive adjective dependent on "city." Gadeira (Phoenician `Gdr, Gadir or Gades in Latin, modern Cádiz) was the primary settlement on the southwestern Iberian coast. It was founded by Phoenicians from Tyre as a trading outpost, allegedly around the time of the Trojan War but more probably in the eighth century BC.[61] It remained a major port city into the Renaissance but today has few visible remnants of its past. The suggestion that it produced great sea monsters (*keteia*, or *kete*) is perhaps less peculiar than it seems. It had (and still has) a vigorous fishing industry, especially tuna, and even though the word can apply to that fish, this seems hardly worthy of note since tuna is common in many places in the Mediterranean. The reference probably refers to large oceanic species such as whales, which were unusual within the Mediterranean but common around Gadeira (located on the Atlantic), and were a matter of curiosity to Greeks.[62] Moreover, there was at least one instance of a large sea creature being cast ashore at Gadeira, whose tail was 16 cubits (about 7 m.) long and which had 120 teeth. This was reported by Turranius Gracilis, a Gadarene from the first century BC who thus was later than the *Periodos*, but the story may have been from an earlier era.[63]

Two days' sail from Gadeira was the prosperous city of Tartessos, located at the mouth of a river that originally had the same name, probably the one later known as the Baitis (modern Guadalquivir). The toponym Tartessos, often applied more generally to this region of southwestern Iberia rather than merely to a single city, was ancient, and probably the same as biblical Tarshish, mentioned regularly in the Bible as a source of King Solomon's wealth. Tartessos came to be a metaphor for the riches of the Iberian peninsula, first exploited by Phoenicians and then by Greeks when Kolaios of Samos reached the region about 630 BC and Tartessian metals were turned into Greek luxury products.[64]

Tartessos was an outlet for the rich mining regions of the interior, exploited until recently as the Río Tinto district. Although the author of the *Periodos* suggested that the resources were alluvial, most were excavated material from the hinterland. Both gold and copper were primary products, but tin probably did not come from Tartessos, but from points farther north, especially the mysterious Kassiterides, or Tin Islands, somewhere in northwestern France or the British Isles. Yet it presumably reached export markets in the Tartessos region.[65]

Since the minerals available at Tartessos were said to be washed down the river, the author considered the interior, to the north. He called this Keltike, but added the peculiar statement that it extended as far as Sardo (Sardinia), which might be literally true but is an odd way of characterizing the Keltic population, since normally it was associated with the Iberian peninsula and mainland regions to the north.

Lines 170–82

The author digressed to discuss the ethnic division of the fringes of the inhabited world and its four major populations. This is material almost certainly taken from Ephorus, who was one of the major sources of the *Periodos* but who is not named in this passage. Similar thoughts were attributed to him by Strabo and Kosmas Indikopleustes, each with its own differences but similar in content.[66]

Ephorus mentioned the four ethnic groups at the margins of the *oikoumene*, one of the earliest statements to consider the world in terms of its overall demography and cultural geography. He set forth his views in the mid-fourth century BC, before geography had evolved into a discipline, and the recension by the author of the *Periodos* is the earliest extant summary of his views on this topic. The Aithiopians were toward the south ("midday"), and the Kelts were in the direction of the *zephyros* (the west wind), the sunset orientation of Strabo and Kosmas. The Skythians were toward the *boreas* wind, or the north (confirmed by the two other sources), and the Indians were in the east ("between the summer and winter sunrises"). Repetition of the Kelts may be poor editing by the author of the *Periodos*; the second entry locates them somewhat more to the west. All three sources agree that the Aithiopians and Skythians were the larger of the four, but this is probably based more on geography than demography, since the east–west length of the *oikoumene* was over twice its north–south width—something taken for granted since the fifth century BC—and thus populations in the north and south would extend over a greater distance.[67] The *Periodos* is the only source to mention the climate of those in the north and south: hot in the south and damp in the north, which assumes (without mentioning it) a temperate region in between. This may be Ephorus' own reflection of the zonal theory of the earth established by Parmenides of Elea in the fifth century BC, who divided the earth into five zones (doubling the temperate and cold, or damp, ones) based on their climate.[68]

Lines 183–95

Some general comments about the Kelts follow. As noted, they were believed to occupy the northwestern quadrant of the inhabited world and were the indigenous population of that region. To the author of the *Periodos* they extended from the upper Istros (Danube) to the Atlantic, but the ethnym was obviously generic and did not take into account any local differences between the various population groups. Greeks had known about them from at least the fifth century BC. Some, at least, were open to hellenization at an early date, especially after Greek traders and merchants from Massalia moved up the Rhone into the Keltic territory, and by the fourth century BC their phil-hellenic qualities were a matter of record.[69] The civilizing power of music was a normal Greek point of view—as ancient as Orpheus—and attributing it to a non-Greek ethnic group only reinforced their perceived superior qualities.

The Column of Baria remains a problem, as no such feature is otherwise known. Various emendations have been suggested.[70] Assuming that "Baria" is incorrect (not a certainty), the best possibility is "Boreios," or North Wind, since the column was the farthest point in the Keltic territory and a pro-montory extending into the External Ocean. This would suggest the north-west corner of France, in the region of modern Ushant, or even a western point in the British Isles (interestingly the region of tin production). A Column of Boreios need not be a human construction but could well be a prominent topographical feature, and the name would be remindful of the Hyperboreians, the semi-mythical people of the far north. Columns at the extremity of the known world, natural or human constructions, were a common feature: in addition to the Pillars of Herakles at the west there were the Pillars of Dionysos or Herakles at the eastern end.[71]

What seems a rather sudden change from the extreme northwest to the upper Adriatic is merely a citation of peoples who were styled the "Farthest" (Eschatoi). The passage is not grammatically coherent, but this does not affect its meaning. The Enetians (the name is preserved in the modern Vene-tians) lived at the head of the Adria (Adriatic). They were perhaps an unusual choice to consider among the farthest peoples of the inhabited world, but it was believed that they had migrated from northwestern France, the presumed location of the Column of Baria.[72] The upper course of the Istros (Danube) was still uncertain at the time of the writing of the *Periodos*, and there was a persistent theory that the river was somehow connected to the Adriatic. This was in part due to the closeness of streams flowing into the sea and the river system (a separation of only about 20 km.), as well as the tale, popularized by Apollonios of Rhodes in the third century BC, that the Argonauts had gone up the Istros and eventually reached the Adriatic. The actual upper course of the Istros (called the Danuvius) was not discovered until Roman penetration of the region north of the Alps in the late first century BC.[73]

Lines 196–216

Returning to the Iberian coast, the *Periodos* summarized the ethnic history of the southern and eastern parts of the peninsula. The term Libyphoenician was a hellenized version of how the Carthaginians described themselves, as at the beginning of Hanno's account. The author of the *Periodos* used it for the Carthaginian settlements in southeastern Iberia, from the Pillars of Herakles around to the north, including such localities as Malaka (modern Málaga) and New Carthage (modern Cartagena). But the three ethnic groups, Liby-phoenician, Tartessian, and Iberian, were presented in reverse chronological order: the indigenous Iberians were the earliest, followed by the Tartessians, who moved into the region from the west, and then the first Carthaginian outposts in the third century BC.[74] Greek settlements, which extended through this region as far as Mainake (line 147), were not mentioned.

The Bebrykians lived in the northeastern part of the Iberian peninsula and were an ethnic group noted for their primitive nature. They were herdsmen who produced milk and cheese, living on the Tyrius (modern Turia) River, which flows into the Mediterranean at modern València.[75] They were at the northern limit of the ethnic groups previously mentioned, where settlement patterns transitioned to the major region of Greek and Ligyan (Ligurian) populations.

Massalia (modern Marseille) was founded around 600 BC, as Timaios reported, by Phokaian settlers from Asia Minor, and located near the mouth of the Rhodanos (modern Rhone) River.[76] This allowed them access to the Keltic hinterland. It quickly became the most important Greek city in the western Mediterranean. Its impressive situation is still apparent today, enhanced by numerous visible remains. Before long the Massalians created settlements along the coast both to the west and east, mixing with the indigenous Ligyan population, as summarized in the *Periodos*. Some of the towns may originally have been Phokaian, but were soon assimilated by Massalian expansionism.

The first major town along the Iberian coast beyond the Carthaginian region was Emporion (modern Ampurias or Empúrias in Spain), whose name reveals its role as a trading post connected to the interior, perhaps originally the farthest Phokaian outpost.[77] To its north, across the Bay of Rosas, was Rhode (modern Rosas), which may have been a Rhodian settlement, as suggested in the *Periodos*. Although this remains disputed, the Rhodians were said to have had an early presence in the western Mediterranean, but they withdrew and Rhode eventually came under Phokaian or Massalian control. The reference to a Rhodian thalassocracy refers to the period of extensive Rhodian seapower during much of the Hellenistic period.[78]

Agathe (modern Agde) lies to the west of the Rhone estuary, and Rhoda-nousia, although not certainly located, may have been at the village of Espeyran a few kilometers west-southwest of Arles.[79] The Rhodanos was a major access route to the interior of Europe, and control of trade along it was the basis of much of the Massalian prosperity: on its upper reaches relatively short portages led to the Loire system and the Atlantic, as well as to the Rhine and central Europe.

Timaios' date of 600 BC for the foundation of Massalia is supported archaeologically.[80] Their outposts continued to the east of the city: Tauroeis (or Tauroention) was near modern Le Brusc southeast of Massalia, and Olbia, a rich archaeological site, was near modern Hyéres. Antipolis (modern Antibes, with fine remains) is about 100 km. northeast along the coast. But the author of the *Periodos* was incorrect in saying that it was the farthest Massalian settlement, since this was in fact Nikaia (modern Nice), a short distance to the east.

Lines 217–35

The account continues along the Mediterranean coast into Italy, passing through Ligystine (Liguria) into the Pelasgian territory. Ligystine is a rare form of the toponym, which is more commonly Ligystike.[81] The Pelasgians, whom the author located in the Tyrrhenian (Etruscan) territory, were a mysterious early population found in a number of places in the Greek and Italian worlds. The source here is probably Ephorus. They were mentioned by Homer, who put them on Crete and in the Troad, but they were also found throughout the Greek peninsula.[82] They were said to be the founders of Agylla, the predecessor of the city of Caere (modern Cerveteri), which was eventually taken over by the Tyrrhenians, with the name changed. The story of Etruscan settlement from Lydia due to the activities of Tyrrhenos the son of Atys had been the standard Greek version since the fifth century BC. The stopover among the Ombrikoi (Umbrians) was also part of the account, with the implication that the Etruscan territory (west of the Tiber) had been Umbrian before the arrival of the Lydians. There was certainly constant interaction between the Etruscans and the Umbrians, who adjoined them east of the Tiber.[83]

A digression on the islands west of the Italian peninsula follows. Kyrnos is Corsica and Sardo is Sardinia, which was the second largest island in the Mediterranean, although the difference in size between it and Sikelia (Sicily) is slight, and it was often uncertain which was the larger.[84] The Seirenides, the Islands of the Sirens, are generally thought to be the modern Li Galli ("Roosters"), three islets just offshore of the Sorrento Promontory, the long peninsula that forms the south side of the great caldera that is the Bay of Naples.

The Island of Kirke is an oddity in the list. Usually called Kirkaion (modern Monte Circeo), there is no doubt about its location, which is where the Italian coast south of Rome turns sharply to the east. It had been an isolated mountain, 541 m. high, since the late fourth century BC, when Theophrastos, who saw its unusual flora as proof that it was the home of the sorceress Kirke, noted that it had recently been joined to the mainland.[85] The author of the *Periodos* may have been writing casually, and thinking of Homer, when he called it an island.

The Ombrikoi were above the Pelasgians (who at one time had occupied the Etruscan territory west of the Tiber), meaning merely that they were more inland (east of the Tiber and in the Apennine uplands). The sudden shift to the Latins and the lack of grammatical coherence in the text has led to the assumption of a lacuna; in fact the ethnym Latinoi must be added to make any sense, and a full line is probably missing. Latinos as the son of Kirke and Odysseus was documented by Hesiod, although he became king of the Tyrrhenians, not the Latins.[86] The Ausones were generally said to be an early population of southern Italy, centered in Campania, but the genealogy seems no earlier than the late Hellenistic period.[87]

The eulogistic comment on Rome, even suggesting pretensions toward ruling the entire *oikoumene*, is an early Greek example of such sentiments. A similar point of view was expressed by Polybios at roughly the same time.[88] This attitude may reflect the sudden eruption of Roman power, especially in the eastern Mediterranean, after the Pergamene legacy of 133 BC—shortly before the composition of the *Periodos*—when Rome gained extensive territory on the continent of Asia; thus the Republic controlled a region from the eastern Aegean west to the Iberian peninsula. Romulus (with variants to his name) had been consistently identified since the fourth century BC as the founder of Rome, or closely associated with the event.[89]

Lines 236–53

The description moves into southern Italy. The Opikoi, more generally known as the Oscans, were an indigenous population around the Bay of Naples, thought to be the same as the Ausonians.[90] Lake Aornos (modern Lago d'Averno) is a volcanic crater, still prominent west of Naples. To its northwest is the ancient city of Kyme (Latin Cumae, modern Cuma), with its spectacular remains. It was founded in the late eighth century BC and was the oldest Greek city in Italy, settled from Chalkis on Euboia.[91] The author of the *Periodos* is the only source to have the Aiolians (in northwest Asia Minor) join the settlement, which may be a detail provided by Ephorus, who was from Kyme in Aiolis. There were connections between this city and Central Greece: the best-known example is Hesiod's father, who emigrated from the former to the latter.[92]

The Kerberion, a place associated with the mythical three-headed dog Kerberos, would have been an entrance to the underworld. The region of the Bay of Naples, with its conspicuous vulcanism, was a natural place for such a locale, including the famous oracle of the Sibyl at Cumae. Kerberos, however, was especially recognized through his image on the local coinage.[93] Association of Odysseus with this region is post-Homeric and probably late Hellenistic: the *Periodos* is the earliest extant author to make such a connection.

Neapolis (modern Naples), which lies about 15 km. east of Kyme, was said to have been founded by a variety of Greek states, perhaps as early as the seventh century BC. Originally there were probably several individual villages, which in time consolidated under the name New City.[94] Kymaians are usually not included among the list of settlers, but there is little reason to doubt their involvement, and according to Strabo, "certain Kampanians" were among the late arrivals. He also cited the role of an unspecified oracle, something almost inevitable in the foundation of Greek cities.

There is a catalogue of ethnic groups in southern Italy, listed roughly north to south. The Ausones were mentioned at line 228; the Saunitai (known to the Romans as the Samnites) lived in the uplands north and east of the coastal plains, and were in the process of being assimilated into the Roman system when the *Periodos* was being written.[95] The Leukanoi, or Lucanians, were to their south, and were related to the Saunitai; they too were losing their independent identity.[96] The Kampanoi (Campanians) were the indigenous population of the plains of Campania, located to the interior of Naples and in one of the most fertile regions of Italy, which today is rapidly disappearing because of urbanization. Farther to the south were the Oinotrioi, whose Greek-sounding name (allegedly they were descendants of Oinotros, who arrived long before the Trojan War) may have supported a persistent idea that this was a region of Greek settlement.[97]

Poseidonias (a form used by the author for metrical reasons; the toponym is Posidonia), a spectacular archaeological site today (Latin and modern Paestum), is located at the southern end of the Gulf of Salerno. It was founded by citizens of Sybaris (for which, see lines 337–60) in the sixth century BC.[98] Elea (Velia) was founded by a Phokaian expedition led by a certain Kreontiades in the 540s BC. It was most famous as the home of the polymath Parmenides and his student Zenon.[99] Notable remains are visible on a bluff overlooking the sea. Even today one can be astounded at the far-flung nature of the settlements established by the small Ionian city of Phokaia in reaction to the Persian advance. They extended from the Iberian Peninsula to the Black Sea, although the comment has the nature of a witticism.

Lines 254–63

The Islands of Aiolos (or the Liparian Islands, modern Eoli or Lipari) are a volcanically active group off the northeast coast of Sicily. They have long been considered the locale of the Homeric story of Aiolos and the winds.[100] Hiera (modern Volcano) has been the most active; the author of the *Periodos* may have heard about the eruption of 183 BC. Lipara (modern Lipari) is the largest, and the only one regularly settled by Knidians from Karia in Asia Minor in the early sixth century BC.[101]

Lines 264–75

A discussion of Sikelia (Sicily) emphasizes city foundations, a particular interest of the author. It was generally believed that Iberians were among the first settlers of the island, and that it was a place of ethnic and linguistic diversity, the latter described by the neologism *heteroglossa*, a word perhaps first used by Polybios.[102] The name Trinakria is first documented in the fifth century BC, but may be only a variant spelling of the Homeric toponym Thrinakia. The author of the *Periodos* reported that it was Iberian, but the more common explanation is that it is a descriptive reference to the triangular shape of the island.[103] Sikelos was the son of Italos and allegedly migrated with his people from Italy to the island 80 years before the Trojan War.[104] Greek settlers were said to have come 10 generations after the war (perhaps 734 BC), from Chalkis on Euboia, founding Sicilian Naxos on the south-eastern coast, the oldest Greek city on the island.[105] The supposed Athenian origin of Theokles (not documented before the fourth century BC) is suspicious, and may be an attempt to involve them in the overseas expansionism of the Greek world, events from which they were singularly absent.[106] But Naxos seems to have attracted an unusually wide range of settlers, of both Dorian and Ionian ethnicity.

Lines 276–99

There is a list of 15 or 16 Sicilian cities, with a brief mention of their foundation history. It is roughly in chronological order, from Naxos to Kamarina. Many of the settlements were Chalkidean, not actually founded by Chalkis on Euboia, but secondary ones by Naxos (which may have been the only actual Chalkidean town on the island), and then further cities established by these outposts.[107]

Hybla (usually Hyblaia, or just Megara) lies on the east coast of the island, and was created when the local dynast, Hyblon, invited Megarians from Central Greece to his territory in order to counteract the growing power of Syracuse, a short distance to the south. The date was in the early 720s BC.[108] Epizephyrion is somewhat out of place, since it is on the east coast of southern Italy (although since the name is generic—Toward the West—there may be an otherwise unknown settlement on Sicily). The Italian town was founded by Lokrians from Central Greece in the late eighth century BC: it was officially Lokroi (at modern Locri) but came to be called Epizephyrian Lokroi.[109]

Archias was a Corinthian, who, like so many leaders of overseas expeditions, had to leave his home because of difficulties. He eventually founded Syrakousai (Syracuse) in the second half of the eighth century BC.[110] The implication is that he may have been joined by settlers from several Greek cities. Leontine (the name is that of a district; the town is Leontinoi, northwest of Syracuse) was a region claimed by the Naxians, again perhaps in reaction to Syracuse, but the town does not seem to have survived and was abandoned by Hellenistic times.[111] For Rhegion, on the mainland opposite Sicily, see lines 309–12. Zankle was the original name of Messene (a detail seemingly not known to the author of the *Periodos*). It was located at the northeastern corner of Sicily, at modern Messina, and was probably an outpost of Rhegion and an attempt to control the straits between the island and the mainland.[112] Katane (modern Catania) was midway along the east coast and a Naxian outpost, established a few years after that city.[113] Kallipolis, another Naxian settlement, has not been located.[114]

Euboia also has not been found. Mylai (modern Milazzo) was on the northern coast of the island. Himera (modern Termini Imerese) is to its west and was one of its outposts. Tauromenion is modern Taormina, on the northeast coast. To say that all these towns were Chalkidean is somewhat of a simplification, but nevertheless the author has provided a detailed account of Greek settlement on the east coast of Sicily and around to the north, beginning with Naxos and Zankle, and their secondary settlements. The catalogue, although thorough, is purely toponyms with no information about the reason for the foundations. The author provided no sources, but they would have included Thucydides, Ephorus, and especially Timaios, who came from Tauromenion.

To conclude his catalogue, the author added four additional cities outside the region previously discussed. Selinous (modern Selinunte) and Akragas (modern Agrigento), about 75 km. apart on the western portion of the southern coast, are the two most spectacular archaeological sites on the island. Selinous, the more western, was settled from Sicilian Megara Hyblaia, and was the western limit of Greek occupation.[115] Akragas was founded in the early sixth century BC.[116] It was a settlement from the city of Gela, to its east and oddly not mentioned in the *Periodos* although an important town in its own right; it may have been abandoned well before the *Periodos* was written. Kamarina (modern Camarina) was, as noted, founded by Syracuse at the beginning of the sixth century BC but revolted in 552 BC and was destroyed.[117]

There were indigenous ("barbarian") populations in the interior, and the western part of the island came under Carthaginian control by two centuries later when Phoenician settlements of the late eighth century BC (at roughly the same time as the first Greek cities) had evolved into Carthaginian ones. The dividing line ran from west of Selinous northerly to west of Himera. But the Greek cities on the south coast were regularly subject to incursions from the Carthaginians until they were driven from the island at the time of the First Punic War.[118]

Lines 300–17

The account returns to the mainland of Italy. At this point the author used "Italia" in its early sense to mean only the southern part of the peninsula, where the toponym originated.[119] The name began to migrate north when the Romans moved into the region. For Oinotria, see line 247 (as the ethnym).

"Mixed barbarians" means nothing more than several indigenous ethnic groups inhabited the region.

Great Hellas (more familiar in the Latin form, Magna Graecia) was a phrase that was in use by the fourth century BC, but which may have had earlier Pythagorean origins, and which referred to the region of Greek settlement in southern Italy.[120] Terina (probably at modern Nocera Terinese), on the west coast at the narrowest part of Italy, was an outpost of Kroton (see lines 323–5) but of little importance.[121]

Hipponion (at modern Vibo Valentia) lies inland south of Terina. Its history before the fifth century BC is not recorded other than the notice in the *Periodos* that it was founded by the Lokrians, presumably those in Italy.[122] Medma (near modern Rosarno), about 25 km. to the south in a large plain, is also little known.

Continuing down the west coast of Italy, the *Periodos* reaches Rhegion, opposite Sicily (modern Reggio di Calabria). The town seems to have been a joint foundation in the late eighth century BC of several Greek cities, including Chalkis, and had a long and varied history, remaining important today.[123] Epizephyrian Lokris, also mentioned at line 278, lies about 60 km. away across the toe of Italy at modern Locri, where settlers from their namesake Greek territory arrived at the end of the eighth century BC. The historicity of the lawgiver Zaleukos has been debated since the fourth century BC. It is probable that the Lokrians had a tradition of an early law code, but it is unlikely that it was the first to be written down.[124]

The Lokrians in Greece were unusual in that their territory was divided into three parts. The westernmost were the Hesperian or Ozolian, located on the Gulf of Corinth. To their east was Phokis, and then beyond Phokis were the Epiknemidian (from Mt. Knemis) Lokrians and the Opountian (from the city of Opous) Lokrians. There was, of course, also Epizephyrian Lokris in Italy, settled from the various Lokrian populations in Central Greece.[125]

Lines 318–36

The *Periodos* continues north along the east coast of the toe of Italy. Kaulonia (near modern Punta Stilo) is about 45 km. northeast of Lokroi. There may have been Krotoniate settlers at the site, but the founder was Typhon of Aigion in Achaia, probably in the seventh century BC. The linguistic explanation for the toponym, from *aulon*, or hollow, is certainly plausible, especially since the town was said to have originally been called Aulonia.[126]

About 100 km. farther is Kroton (at modern Crotone), founded near the end of the eighth century BC. It is little known archaeologically, yet it became the most important city of the region, in part due to the presence of Pythagoras of Samos, who dominated its cultural presence in the second half of the sixth century BC. Its founder, Myskelos from Rhypes in Achaia, was another example of someone who ran into difficulty at home and was encouraged to emigrate; his career became an almost mythic tale.[127]

Pandosia, another Achaian foundation, was at a location that is unknown today, presumably north of Kroton. Its name ("Giver of Everything") suggests a site noted for its agricultural abundance. The author of the *Periodos* is in error in saying that Thourioi was an Achaian settlement: only its predecessor, Sybaris, was Achaian. Thourioi itself was a pan-Hellenic foundation of 443 BC, established at or near the destroyed site of Sybaris, in the rich floodplain of the modern Crati and Coscile Rivers. It included Herodotus and Protagoras among its citizens, but never prospered.[128]

Metapontion (modern Metaponto) is farther along the coast in its own floodplain. Geographically it was the northernmost Achaian settlement in this region, probably from the late eighth century BC. It was famous as the final home of Pythagoras but was depopulated by the third century BC, possibly due to malaria.[129]

Reserving discussion of Sybaris for lines 337–60, the author examined Taras (modern Taranto), at the head of the gulf of the same name that divides the toe and heel of Italy. In stating that it was the largest city in Italia the author was using the ancient definition of the region, limiting it to the southern part of the peninsula. It was a rare example of a Lakedaimonian settlement, and the eponymous founder, Taras, appears on its coinage as early as the fifth century BC. The Partheniai ("Children of Maidens") were those of uncertain fathers born while the men were away for extended periods, such as at war, and would be a natural contingent for overseas settlement. Taras, still an important city today, is in an impressive location, since harbors are a rarity in southern Italy.[130]

Lines 337–60

The author of the *Periodos* devoted an extended section to the history of Sybaris, the most notorious city in southeastern Italy, whose name produced the word "sybarite." The account is out of place geographically, since Sybaris was near the future site of Thourioi, but this is presumably because the author did not want to interrupt his catalogue of south Italian cities with the lengthy digression.

The city was located in a rich and fertile flood plain, but became a formulaic tale of excess and an indulgent lifestyle. It was founded around 730 BC.[131] Even though its collapse was conventionally said to be due to internal issues, such as lawlessness, flooding and malaria probably also played a role. Competition with Kroton was certainly a factor, and the establishment of their own games— Kroton was a famous athletic center—could be seen as a provocation. In time the Krotoniates gained the upper hand and destroyed the city in 510 BC.

Lines 361–8

Continuing to use Italia in its original limited sense, the author of the *Periodos* moved to the southeastern corner of the peninsula. The Ionian Sea was an ancient name for the entire Adriatic, but later became limited to the portion southeast of Italy and west of the Greek peninsula.[132] The Iapygians were the indigenous population of the heel of Italy and the Messapians were one of their local groups.[133] Brentesion (modern Brindisi) was the major city of the region, notable for its fine harbor and as the primary crossing point to Greece, a role that it still plays. Little was known about its origin.[134] The Keraunian Mountains (modern Karaburun) are actually across the Adriatic in modern Albania, but are visible from the Italian side.

The Ombrikoi (Umbrians) had been discussed at lines 220–6. But line 366 is corrupt and locating them "toward the west" makes little sense: it is not certain that this is the proper emendation. Their traditional region was in central Italy, northeast of Rome, but at one time they did have a presence on the Adriatic to the north of the Messapians. Nevertheless their luxurious lifestyle was noted as early as the fourth century BC, which may in part be confusion or association with the Etruscans, who had Lydian connections. The source for the *Periodos* was probably Theopompos, mentioned in line 370.[135]

Lines 369–97 (401)

The *Periodos* turns to a discussion of the Adriatic Sea and the regions on its coast, but avoids any further examination of the rest of Italy between the Iapygian territory and the northern end of the Adriatic. In one of the few citations within the text of a source, the author named Theopompos of Chios, who was probably used for the following details on the Adriatic as well as material on southern Italy.[136] Theopompos wrote a variety of historical works during the fourth century BC, including a detailed study of the environment of Philip II of Macedonia.

The context is pre-Roman, given the reference to the tens of thousands of barbarians that surround the Adrian Sea. The Adria (also Hadria, modern Adriatic) was known from early times as part of a trade route from central and northern Europe to the Mediterranean, used especially for the transmission of *elektron*, or amber.[137]

Greek geographers were fond of seeing landmasses as isthmuses, not only the obvious ones such as that of Corinth, or between the Mediterranean and the Red Sea, but in many other places less understandable today. Practically anywhere that land was between two seas—an inevitable situation in the Mediterranean world—was considered an isthmus, even if the distance were great, such as that between the Adriatic and the Pontos (Black Sea), whose narrowest stretch, between northern Albania and southern Bulgaria, is still several hundred kilometers. The islands mentioned are all on the northeast coast of the Adriatic, including the Apsyrtides (modern Cres and Lošinj) and the Libyrnides, a general term for the extensive chain along the modern Croatian coast. The comparison with the Kyklades (in the Aegean) is in terms of the density of small islands.

The Elektrides (Amber) Islands are problematic. They were allegedly off the mouth of the Padus (modern Po), where there are no islands other than those formed by the channels of the river. But they are part of the story of Phaethon and were also associated with trade routes from northern Europe. As such the islands are totally mythical, but their name reflects not only their position on the amber route but the existence of amber workings (at least in the Late Bronze Age) in the Po delta.[138]

The figure of 1.5 million barbarians is not documented elsewhere and may be from Theopompos, but there is no doubt that there were fertile regions around the upper Adriatic, including the extensive plain of the Po. The ability of animals to have twins is one indication of fecundity: it was said that the local cattle usually had twins, and goats three or four offspring, or even five.[139] The implication, although not specifically expressed, is that this was due to the climate, which differed significantly on the Adriatic and on the coast of the Black Sea (which are not as close to each other as the author seems to have believed). Some of the meterological phenomena discussed were described by Aristotle.[140]

The Enetoi (see also line 193) were the indigenous population at the head of the Adriatic. They were probably Keltic in origin, but there was a tendency to connect them with the Enetians from Paphlagonia (in northern Asia Minor) who were allies of the Trojans, yet this does not seem to have been popularized until the first century BC.[141]

The Eridanos was a mythical river in the west, associated with the story of Phaethon and his theft of the chariot of his father Helios; the river was known as early as Hesiod. As knowledge of Italy increased, the river came to be equated with the Padus (Po) as well as the Elektrides Islands (line 374) and the assumption of amber production in the region.[142] The tree in question (*aigeiros*) is presumably the black poplar (*Populus nigra*), so familiar in the Mediterranean world, but this is not the one that produces the resin that becomes amber, which comes from pine trees. The matter of the locals who dressed in mourning for Phaethon was a popular tale but probably nothing more than a characteristic of costume turned into a Greek aetiology.[143]

Lines 398(394)–414

The next section of the *Periodos* discusses populations from the northern end of the Adriatic and along its eastern coast, as well as inland. The list is almost totally in terms of ethnic groups. The author's sources probably included Timaios and Eratosthenes, both mentioned in line 412.

Istria is the peninsula protruding into the northeastern end of the Adriatic, now divided between Slovenia and Croatia but retaining an Italian cultural outlook. Although there are many islands in the region, and along the coast to the south, there are no known tin sources, and mention of this may be a misplaced reference to the Tin Islands of northwestern Europe.[144] This is not the only time the *Periodos* seems to have been misinformed about the origin of tin: at line 165 it was said to come from Tartessos.

The Ismenoi were presumably in the interior north of the Adriatic, but are otherwise unknown. The Mentores seem to have been somewhat to their south, perhaps in the islands of northern Croatia.[145] The Pelagonoi were also located in Macedonia (line 621), and were a widespread early population. The Libyrnoi were along the eastern Adriatic coast, both on the mainland and in the islands. The Boulinoi were perhaps the Byllionians of Strabo, near modern Hekal in Albania.[146]

The Hyllike Peninsula is modern Punta Planka, a promontory northwest of Split in Croatia. It is the place on this coast that protrudes the farthest into the Adriatic, but is hardly comparable in size to the Peloponnesos, and it is baffling that such an equation should be made, although comparison of remote areas to the Peloponnesos was a feature of Greek geographical thought. The account is a typical foundation myth oriented on Herakles, including the difficult assimilation of the locals into a Greek environment. The source is probably Timaios, who had an interest in ethnographic details. Eratosthenes' contribution may have been more topographical.[147] The island of Issa (modern Vis) was a Syracusian foundation of the early fourth century BC, part of the grandiose Adriatic efforts of the tyrant Dionysios I.[148]

Lines 415–43

The account continues along the eastern coast of the Adriatic through the territory known as Illyria, south of Hyllike and as far as the beginning of the Greek peninsula. But the term was generic and included large numbers of different ethnic groups. The ethnographical comments are probably from Timaios. The most prominent monarchies in Illyria were that of King Agron in the 230s BC, and that of his wife and successor Teuta. She was an important figure in the global politics of her era, including the first local contacts with Rome. The comments on Illyrian cultural qualities may be formulaic, but also reflect the assumed benefits of the climate.[149]

Pharos Island (modern Hvar) is east of Issa (line 413). The settlement on it was founded in the early fourth century BC from the Aegean island of Paros and was one of a number of Greek cities in the region.[150] Black Korkyra (Korkyra Melaina, modern Korčula) was so called because of its dark forests and also to distinguish it from the better-known Korkyra (modern Corfu) to the south. It was a foundation of Knidos, in Karia in Asia Minor, perhaps around 600 BC.[151] Lake Lychnitis (or Lychnidos) is modern Lake Ohrid on the Macedonian–Albanian border, far inland and somewhat out of place in the catalogue of coastal locations; the author may have wanted to make the comment about Diomedes. There were numerous tales about the hero, who was from Argos and was active in northwest Greece before the Trojan War. After the war he found that there was no place for him at home, and, according to the most common version of his later life, he ended up in Italy and was associated with many locations throughout the peninsula. His tales center on the islands named after him (modern Isole Tremiti) off the southeast coast, where he died.[152] One of them may be the island mentioned in the *Periodos*, but if so it is badly misplaced.

The Brygoi were an inland population which became notorious for attacking the Persian army of Mardonios in 492 BC.[153] Epidamnos was on the coast near the Brygian territory (at modern Durrës in Albania). It was a joint foundation in the late seventh century BC of Korkyra (modern Corfu) and Corinth.[154] The Encheleioi (Eel People) lived around Lake Lychnitis, and, as expected, were notable for their fishing industry. Kadmos, the founder of Boiotian Thebes, came to this region and was turned into a snake, but this event is generally placed on the coast rather than at the inland lake.[155] Apollonia (modern Pojan) was another joint Korkyraian and Corinthian settlement, founded somewhat later than Epidamnos. Orikos, or Orikon (at modern Orikumi) is little known beyond having been a Euboian foundation.[156]

Lines 444–69

The account moves down the west coast of the northern Greek peninsula. The Thesprotoi were in northwest Greece and had been a prosperous population in Homeric times.[157] The Chaones were on the southern coast of modern Albania and were involved in the Peloponnesian War.[158] Offshore from them is the island of Korkyra (or Kerkyra, modern Corfu), the earliest Greek settlement on the Adriatic, originally Eretrian and then Corinthian.[159] Inland, east of the Thesprotoi, were the Molottoi (usually Molossoi), the most powerful indigenous population of the northwestern Greek peninsula. They reached their peak in the fourth and third centuries BC, after the daughter of King Neoptolemos, Olympias, married Philip II of Macedonia and the couple became the parents of Alexander the Great. Another famous member of the dynasty was Pyrrhos, Olympias' cousin and noted for his invasion of Italy in the early third century BC. But the author of the *Periodos* was in error in reporting that he was the son of Neoptolemos; his father was actually Aiakides, Neoptolemos' nephew.[160] The oracle of Zeus at Dodone (or Dodona), probably the most famous in the Greek world before the rise of Delphi in the sixth century BC, was in Molottian territory and an important element of their prestige.[161]

The narrative returns to regions closer to the coast. Ambrakia (at modern Arta) was an inland city, but well connected to the sea through the Aratthos River and the Ambrakian Gulf. It was a Corinthian settlement of the late seventh century BC. Gorgos was the second (not the eldest) son of the tyrant Kypselos, and may have wished to leave Corinth when his older brother Periandros succeeded their father in the 620s BC.[162] The location of Amphilochian Argos is not known, but it was probably southeast of Ambrakia, on the Ambrakian Gulf. Most accounts attribute its foundation to another son of Amphiaraos, Alkmaion (or Alkmeon), whose activities were described in the epic known as the *Alkmaionis*.[163] Both brothers, Amphilochos and Alkmaion, could have been involved in establishing the city.

Anaktorion was also on the Ambrakian Gulf (near modern Nea Kamarina). It was essentially a Corinthian foundation, but people from nearby Akarnania could have migrated to the new city. Akarnania itself is a district, not a town, the region extending from Anaktorion and the south side of the Ambrakian Gulf along the coast to the outlet of the Gulf of Corinth. Greek settlement in this area by Alkmaion's son Akarnan was described in the *Alkmaionis*.[164]

The *Periodos* lists the major islands on the west coast of the Greek peninsula. Leukas (modern Leukas or Leucadia) is the northernmost, lying just a few hundred meters from the Akarnanian coast. Its major city, where the Corinthians would have settled, is at the northeast end, and was established during the tyranny of Kypselos in the mid-seventh century BC.[165] Kephallenia (modern Kephalonia) is about 15 km. south of Leukas. The author of the *Periodos* used the Homeric convention of describing it only by the ethnym of its inhabitants.[166] Ithake (Ithaka) is off its northeast coast.[167] Zakynthos is about 15 km. south of Kephallenia and 20 km. west of the westernmost promontory of the Peloponnesos.[168] Returning north, the author mentioned the Echinades, a group lying off the southwest coast of Akarnania and northwest of the mouth of the Acheloos River. This is a region where extensive siltation has continually altered the topography since earliest times.[169]

Lines 470–87

At this point the *Periodos* begins a lengthy description of the central and southern Greek peninsula, one of the longer sections of the text (through line 617). Acknowledging his debt to Ephorus for the account of all of Greece, the author began with the territories on the north side of the Gulf of Corinth. Aitolia was to the east of Akarnania and the Acheloos River, and was settled by people moving across the gulf from Elis, under the leadership of an eponymous Aitolos: Ephorus cited epigraphical information to document this.[170] The Kouretes were the indigenous population of the region and were forced out when the Eleians arrived. Naupaktos is on the north side of the Gulf of Corinth, just east of Rhion. It was believed that the Dorians entered the Peloponnesos here in one of the major population movements after the end of the Bronze Age, and Rhion always was, and remains, the major crossing point, although the city of Naupaktos seems to have been of little importance before the fifth century BC.[171] Temenos was the Dorian leader, whose traces were found throughout the Peloponnesos and who was the subject of Euripides' play *Temenos*, which discussed the populating of the region.[172] Rhion was a topographical point, not a town, at the north side of the narrowest point of the Gulf of Corinth.

The Ozolian Lokrians, already mentioned at line 317 in connection with Lokroi in Italy, were the westernmost of the Lokrian groups, which were widespread across Central Greece. The origin of their specific name, Ozoloi (perhaps "Strong Smelling") is obscure.[173] The Lokrians "turned toward Euboia" are the other two groups, the Opountian and Epiknemidian. None of the Lokrians ever controlled Delphi or its oracle, as far as is known (it was in Phokian territory), but since the eastern boundary of Ozolian Lokris was only a short distance from Delphi, there may have been a local claim. Phokis, to the east, was between the groups of Lokrians. The genealogy of the eponym Phokos is generally accepted, from Sisyphos through his son Ornytos (or Ornython) to his son Phokos: these were all from an early Corinthian dynasty.[174] It is reasonable that Corinthians migrated across the gulf and settled in Phokis.

Lines 488–501

The account of Boiotia is scant but emphasizes its unique position with harbors facing both west and east. Those on the south side (toward "midday") are on the Gulf of Corinth at locations such as Siphai and Kreusis, and provided access to the Adriatic and Sicily. To the east, on the southern part of the Euboian Strait (which separated Euboia from the mainland) were Aulis and Delion, from which one could go to Cyprus, Egypt, and the Aegean islands. At the north end of the Euboian Strait Anthedon and other ports were the beginning of routes to Macedonia and Thessaly.

The author of the *Periplous* named four Boiotian cities in addition to the small seaport of Aulis, famous for where the Achaians gathered on the way to Troy. The site of Tanagra (the text uses the ethnym) lies about 15 km. south of Aulis at modern Grimadha, located at the interior edge of the coastal plain. In Hellenistic times it was the most important city in eastern Boiotia.[175] The other major city of the era was Thespiai (at modern Erimokastro), 35 km. to the west of Tanagra. Anthedon (at modern Loukisia) is on the northern coast, and Thebai, or Thebes, was the most famous city in Boiotia, with a rich mythological history, and remaining important into Classical times. Yet after its destruction by Philip II and Alexander in September 335 BC it never regained its prominence and lived on its historic reputation.[176]

Lines 502–34

Continuing around the head of the Corinthian Gulf, the *Periodos* mentions the Megarian territory, which lies between Boiotia, the Corinthia, and Attika. There was Bronze Age habitation in the region, but the city of Megara was probably not founded until the eighth century BC, not long before it sent out settlers to Megara Hyblaia in Sicily (line 277). The eponymous Megareus was probably Boiotian in origin (Onchestos was a town near Thebes), and it is reasonable that some Boiotians might seek a presence on the Saronic Gulf. Megareus was said to have fought King Minos during the brief Cretan involvement in the region.[177] Yet it is unlikely that settlers from Messene (in the southwest Peloponnesos) came to Megara, and there may be a confusion with Messana in Sicily.

The account moves west into the Peloponnesos. The Kenchrean Gulf is a rare term for the innermost part of the Saronic Gulf at the Isthmos of Corinth and near the town of Kenchreai, the eastern port of Corinth. The rugged and indented coastline of the Peloponnesos was a well-known hazard to shipping, especially in its southern portions. Cape Malea (or Maleai) was its southeastern point, where ships needed to turn in order to go from the Aegean into the Mediterranean proper. It was notorious for navigational difficulties and shipwrecks, beginning with Odysseus and Menelaos.[178] Cape Tainaron, farther west, is the southernmost point of the Peloponnesos and also an important sailing point where ships would turn toward the north to access the western Peloponnesos and the Adriatic. Its Sanctuary of Poseidon, a Lakonian establishment, was an early shrine with notable art work, almost certainly visible from the sea.[179]

The ethnic summary of the Peloponnesos, probably from Ephorus, is succinct but thorough. But Ephorus was probably not the only source, since there is an emphasis on navigational details, including winds. A possibility for the second source is Timosthenes of Rhodes and his nautical guide for Ptolemaic seamen, used by the author of the *Periodos* (line 118).

The account of the Peloponnesos moves around in a counter-clockwise direction from the region of Corinth, with Sikyon to its west, and then Achaia, all on the north side and thus in the direction of the *boreas* (north) wind. On the west side were the Eleians to the north and the Messenians to the south. Their location is expressed by a tautology, toward the *hespera* and *zephyros* (the evening and the west wind), perhaps indicative of two sources used. The location of the Lakonians is also described in two ways: toward midday, a conventional way of indicating a southern direction (see line 491) and toward the southern *klima*. This is a technical term meaning a terrestrial zone or latitude, probably the invention of Eudoxos of Knidos in the fourth century BC, another indication that a source in addition to Ephorus was used.[180] To place the region of the Argives in the south only applies from the Corinthia, and suggests that one of the sources for the *Periodos* was oriented on that area. Akte (see also line 533), or "Promontory" refers to the eastern ("toward the sunrise") projection of the Argive region, around Epidauros.[181] There is no obvious reason why Phleiasia, the district of Phleious in the Corinthia, should be singled out. Its sole claim to fame is that it was said to be the home of the family of Pythagoras.[182] Arkadia was the only land-locked district of the Peloponnesos, and indeed of central and southern Greece.

The founders of the major regions of the Peloponnesos are listed in a catalogue taken from Ephorus.[183] Only the interior Arkadians escaped invasion by outside groups. Aletes ("The Wanderer") ruled Corinth for 38 years.[184] Phalkes was a son of Temenos (line 479), who was the leader of the group invading the Peloponnesos. Tisamenos moved around the Peloponnesos but ended up in Achaia.[185] Oxylos came to Elis two generations after the Trojan War (which provides a general chronological datum for the invasions), and under his leadership the district prospered.[186] Kresphontes was notorious for using trickery in order to obtain Messenia.[187] Eurysthenes and Prokles established themselves in Lakedaimon and were the ancestors of the dual Spartan monarchy.[188] Kissos (or Keisos) of Argos and Agaios, settling in the eastern Argolid, were sons of Tisamenos.[189] Deiphontes was his brother-in-law. This list of settlers provides an excellent picture of a single clan—perhaps relatively small in number—crossing into the Peloponnesos at Rhion and then spreading its family groups through the territory over the next two generations.

Lines 535–49

Before continuing with the rest of the Greek peninsula, the *Periodos* examines the island of Crete, with at least some of the details taken from Ephorus. There was actually a point of view that saw Crete as an extension of the Peloponnesos, since the latter is the closest mainland to the island.[190] It is the fifth largest island in the Mediterranean and the largest in the Aegean region, creating a geographical barrier to the sea, extending about 360 km. across its southern end, effectively from Cape Maleia to Rhodes, both of which are about 95 km. from the island. Its dense population, with 100 (or 90) cities, had been taken for granted since Homeric times.[191] The first comments on Crete have some affinities with the Cretan story that the disguised Odysseus told Penelope.[192]

The Eteocretans ("True Cretans") were also known to Homer as one of the primitive ethnic groups of the island, but were eventually believed to have been the original settlers.[193] The Cretan thalassocracy under King Minos and the connection with early sea power were well-known elements of the early history of the Aegean world. The eponymous Kres is probably merely a back formation from the name of the island. The landing points for the day's sail are not indicated, but the timing is perfectly reasonable. There are many records of sailing times, which indicate that ships could easily do 100 nautical miles in a day.[194] This rare citation of a sailing time within the *Periodos* is further evidence that the author had access to a sailing manual for his discussion of the Peloponnesos and its environs (see also line 151).

Lines 550–8

Four islands are mentioned, situated from the eastern Aegean to the Saronic Gulf. There is no obvious selection process, or why these should have been singled out. Astypalaia is a small isolated island in the southeastern Aegean, lying about 160 km. northeast of eastern Crete. Citation of Megarian origins suggests that the author of the *Periodos* may have obtained the information from the Megarian source (perhaps Ephorus) that is common in the *Periodos* (see also lines 277, 292). The unusual toponym Cretan Strait is also peculiar for a locality so far from Crete, but the account may have been a vestige of a sailing route northeast from Crete to the eastern Aegean and Asia Minor.

Kythera lies between Crete and the Peloponnesos, about 15 km. from the latter. It was famous as the last landfall of Odysseus before he entered unknown territory.[195] Aigina is in the middle of the Saronic Gulf. It is accepted that it was originally called Oinone or a similar name before the family of Aiakos (the grandfather of Achilles and Aias) took control. At that time it was named for Aiakos' mother Aigine, the daughter of the river god Asopos.[196] The island, famous for its well-preserved Temple of Aphaia, was a major state in pre-Classical Greece. About 15 km. to the north, lying against the coast, is Salamis, one of the most famous islands in Greek history, and also ruled by the family of Aiakos in early times.

Lines 559–65

Returning to the Greek mainland and to Central Greece, where the narrative had been before covering the Peloponnesos and the islands (line 507), the next locality for discussion is Athens. The material is closely derived from Herodotus, as the author made clear.[197] The Pelasgians were the early inhabitants of the Greek peninsula, located in several places (lines 217, 226), with the group called the Kranaoi coming to Athens. The word may be descriptive, meaning (as *kranaos*) "rocky" or "rugged," a reference to the local topography: Athens was called the "rocky city."[198] Kekrops was the first king of Athens and the original civilizing figure of the city, followed by another seminal personality, Erechtheus, who allegedly was the one who named it after the goddess.[199]

Lines 566–86

The narrative follows the coast of Attika, effectively in the style of a *periplous*, but no places are named until Sounion, the southernmost part of the peninsula. Euboia is the long, narrow island along the east coast of Attika and Boiotia, the second largest Aegean island (after Crete). The author has probably returned to Ephorus as a major source, since he seems to have been the first to document the ancient descriptive name, Makris.[200] But in addition the account of Euboia is unusually long for the *Periodos* and may reflect the heavy involvement of its citizens, especially those from Chalkis, in overseas settlement (lines 238, 276, 290, 311), as well as the author's use of a local, Dionysios of Chalkis, as one of his sources (line 116).

The nymph Euboia was one of the daughters of the river god Asopos. The Leleges were another of the scattered early populations of the Greek world.[201] Chalkis, located at the narrows known as the Euripos, where Euboia comes within a few meters of the mainland, was the prominent city on the island. There was a strong Athenian claim to both Chalkis and Eretria (the second most important Euboian town), but it is not well documented.[202] Aiklos and Kothos were brothers of Athenian origin. Kerinthos (modern Mandouli) is on the eastern coast of the island, and was known to Homer.[203] The Dryopes ("Oak People") were another early population located in several places, from Thessaly to the Peloponnesos.[204] Karystos, also known to Homer, was at the southern end of the island and was famous for its quarries.[205] Hestiaia, or Histaia, another Homeric locality, was at the north end of the island.[206] The Perrhaibians were Thessalian in origin and might have been expected to cross over into northern Euboia.

Continuing to rely on a Chalkidean source, presumably Dionysios, the author mentioned four Aegean islands that were eventually settled from Chalkis. They lie immediately northeast of Euboia and are the group called the Northern Sporades.[207] Skyros is the largest and the most important, isolated somewhat to the east. It had a surprisingly rich mythological history and came to be known for its goats and marble. Peparethos (modern Skopelos) was said to have originally been a Minoan outpost, and much later an Athenian one. Ikos (modern Alonnesos) lies across a narrow channel to its east and was of little importance. Much the same can be said for Skiathos, the westernmost of the four and close to the mainland. All of them were said to have had a similar history: habitation by early peoples (Pelasgians or Cretans), and then eventual desertion and repopulation by the Chalkideans. The only personality named is Staphylos, a commander of Rhadmanthys of Crete, who sent an expedition into the Aegean. A total of eight outposts were established, scattered from Maroneia on the Thracian coast south toward Crete, literary evidence of the Minoan thalassocracy.[208]

Lines 587–617

The account continues to the west of Boiotia (see lines 488–501) with the Lokrians. These are the Epiknemidian and Opountian portions, since the more western Ozolians had been discussed at line 481. There is an unusually lengthy genealogy running through five generations. Deukalion, son of Prometheus, was an early civilizing figure in the Greek world; he and his wife Pyrrha were the creators of the human race.[209] Amphiktyon was also an important regional figure. Itonos was connected with the sanctuary of Itonian Athene in western Boiotia and another one in Thessaly, and thus was an early cultic figure, although the person named, in line 590, is corrupt in the manuscript. Physkos is little known, and the eponymous Lokros expelled the indigenous Leleges (for whom, see line 572).

Although the Dorians were seen as one of the major components of Greek ethnicity, there was a small territory west of Lokris and north of Phokis named Doris, from which they were said to have originated. It had associations with Herakles, but otherwise was quite obscure in Greek history. The four cities named, Erineos (modern Evangelistra Kastellia), Boion, Kytinion (modern Palaiochori), and Pindos (modern Ano Kastelli) were a tetrapolis that exercised some control over the access into Central Greece from the north. None of them is more than a few kilometers from each other, although the exact modern locations are not certain. The eponymous Doros may have some legitimacy as an early settler, since he was known to Hesiod.[210]

Herakleia, or Herakleia Trachinia, was in the region of Trachis and a short distance west of Thermopylai. It was founded by the Spartans (Lakones or Lakonians) in 426 BC as an outpost—obviously heavily populated—against Athenian activities in the region.[211] Pylaia was a sanctuary of Demeter ("Demeter of the Gates"), and the Amphiktyony ("Those Living Around") was composed of the locals who were responsible for the maintenance of the sanctuary.[212] These places were on the south side of the Malaic Gulf, the deep bay that separates Central Greece from Thessaly. Echinos (at modern Achinos) is on the north side of the gulf near its outlet, and was one of the cities subject to Achilles; the founder, Echion, was one of the five Sown Men from Thebes who rose up from the teeth of the serpent killed by Kadmos.[213]

The Phthiotic Achaians were the inhabitants of the southern part of Thessaly, the home region of Achilles.[214] The ethnym reflects the early and Homeric usage of Achaia as refering to a wide area of the Greek peninsula. The Magnetes lived in Magnetis (or Magnesia), the peninsula on the Aegean side of Thessaly, which contained Mt. Pelion (1625 m.), famous as the home of the centaur Cheiron.[215] The territory "above" the Magnetes, or inland, is the extensive Thessalian plain, the largest in the Greek peninsula. Larisa, in the northeast, has been the most important regional center since the Archaic period.

The Peneios River flows for 160 km. across northern Thessaly. Its volume is one of the largest in Greece, and it has been an important feature of Greek topography since Homeric times.[216] The river empties into the sea through the Tempe gorge, the "Vale of Tempe" of literature, which extends for about 10 km. near its outlet. Lake Boibeis was a remnant of the prehistoric lake that covered the Thessalian plain, which was said to have been drained by either Poseidon or Herakles, perhaps a memory of early volcanic activity.[217] The area of Boibeis eventually became marshland and was finally totally drained.

There follows a summary of the main ethnic groups in and around Thessaly. Athamania was a small territory in the mountains to the west. It had a brief prominence in Hellenistic times and was involved in the growing Roman presence in the Greek peninsula, with its people eventually becoming assimilated.[218] The Dolopians were southwest of the Thessalian plain; the region was the home of Achilles' mentor Phoinix.[219] The Perrhaibians lived north of the Peneios above Tempe and were the original inhabitants of the region. The Ainianians occupied a small territory between Doris and Dolopia; their variant name, Haimonians, reflects one of the early inhabitants of Thessaly, Haimon the father of Thessalos.[220] The Lapiths were associated with the region above Tempe but are probably mythical, yet famous because of their depiction on the Parthenon metopes. The Myrmidons were the traditional population of Thessaly itself and associated with the world of Peleus and Achilles, mentioned regularly by Homer.

Lines 618–63

Following is a lengthy section on Macedonia, beginning at its southern boundary on the Peneios River. Olympos, at 2917 m., the highest mountain in the Greek peninsula, lies immediately north of the river. The eponymous Makedon was the son of Zeus and the grandson of Deukalion.[221] The Lynkestoi were in the west of Macedonia; their king Arrhabaios was the great-grandfather of Philip II of Macedonia.[222] The Pelagonoi have already been noted as an early population scattered from east of the Adriatic into Macedonia (line 413). The Axios River (modern Vardar in Macedonia) flows south from near Skopje into the Aegean, and has long been the major land route into Greece from the north. The Botteatai were in eastern Macedonia in the Chalkidean region (not the locality on Euboia, but the prominent district that forms three conspicuous peninsulas in eastern Macedonia), and the Strymon River, known to Hesiod, was the original eastern boundary of Macedonia before the expansionism of Philip II.[223]

The list of Macedonian towns follows no detailed geographical order, which may be for metrical reasons.[224] Pella was the Macedonian capital, about 20 km. inland. It came to prominence in the fifth century BC and was the largest town in the territory.[225] Beroia (familiarly Berea) was to its southwest at the edge of the plain.[226] Thettalonike (Thessalonikeia, modern Thessaloniki), today the most important city in the region, was not founded until the late fourth century BC, and as part of the political manoeuvering of the successors of Alexander the Great, in order to give Macedonia a coastal port.[227] Pydna was on the coast in southern Macedonia. It had existed since at least the fifth century BC but was most famous because the Romans defeated Perseus there in 168 BC, bringing the Macedonian kingdom to an end.[228]

Aineia was the western promontory of the Chalkidike peninsula and was probably one of the towns whose population was moved to establish Thettalonike, about 15 km. to the north. Potidaia, at the narrows of the western peninsula of the Chalkidike, was a Corinthian foundation of around 600 BC, and then was reestablished as Kassandreia in the late fourth century BC.[229] Antigoneia was probably near modern Agios Pavlos, about 8 km. inland, and was founded in the late fourth century BC. Olynthos was at the head of the

Toronic Gulf, which lies between the western and middle peninsula of the Chalkidike. It was notoriously destroyed by Philip II in 348 BC.[230]

The narrative continues more or less toward the east from the Olynthia, the region of Olynthos. Arethousa (at modern Rendina) is about 50 km. north-northeast of Olynthos.[231] Pallene is the name normally given to the westernmost peninsula of the Chalkidike, but no town of that specific name is known, although the implication is that any townsite would have been at the north end of the peninsula. The original name, Phlegra ("Inflamed") was applied to the entire region, one of the places where Herakles was believed to have fought the Giants. It was said that ships from Pellene in Peloponnesian Achaia were driven to this place by a storm and settled on the peninsula, giving it the name Pallene.[232]

The Toronic Gulf is between the Pallene and the middle peninsulas of the Chalkidike. Mekyberna (or Mekyperna, at modern Molivopirgos) was at the head of the gulf, near Olynthos, but had been abandoned when the *Periodos* was written, with its population probably moved to assist in creating Kassandreia (line 630) in 316 BC. Torone itself was at the mouth of the gulf on the west side of the middle peninsula.[233]

Mention of the island of Lemnos may seem to be anomalous at this point in the narrative, since it is in the middle of the northern Aegean, but Torone, about 100 km. to its west, was the closest city on the mainland. The island was sacred to Hephaistos, since this is where he landed when he was thrown from Olympos. The mythical settler of the island was Thoas, the son of Dionysos and Ariadne, who was vaguely connected with the Argonauts. In the late sixth century BC the younger Miltiades took over the island for the Athenians, as part of his activities in the northern Aegean.[234]

Athos is the eastern peninsula of the Chalkidike. The reference to sailing demonstrates that one of the sources of the *Periodos* at this point was a *periplous*, perhaps that of Timosthenes of Rhodes (line 118). Akanthos (modern Ierissos) is on the eastern side of the peninsula, and was founded by the island of Andros in the mid-seventh century BC. Just to its east are the remnants of the canal that Xerxes cut through the peninsula, which was about 1.5 km. long.[235]

Amphipolis, about 45 km. north of Akanthos and east of the Chalkidike, lies about 5 km. inland on the left bank of the Strymon River. It was an Athenian foundation of 447 BC which played an important role in the Peloponnesian War, and is a major archaeological site today.[236] The Strymon, flowing south from western modern Bulgaria into the Aegean, was the original eastern boundary of Macedonia.[237] The Dances ("Choroi," line 652) of the Nereids are otherwise unknown and defy a meaningful explanation. The Nereids were sea nymphs and appear throughout Greek myth, but the word *choroi* is not known to have been otherwise used topographically. A possible intepretation is that the feature was rapids or riffles on the lower river.

Berga has not been exactly located but was probably west of Amphipolis. It is difficult to understand why its citizen Antiphanes should be one of the few people mentioned by name in the *Periodos*, but citation of him may have been lifted from whatever source the author was following at this point, perhaps with elimination of a fuller account. Antiphanes was an obscure geographical author whose writings were categorized as fantasies, an easy charge to make against accounts about the fringes of the world. His one known topic was the frozen north, which suggests that he was later than Pytheas of Massalia, who explored the Arctic in the late fourth century BC. Antiphanes was cited by Eratosthenes, so his date must fall between the time of Pytheas and the late second century BC.[238]

Emathia was believed to be Homeric Oisyme, or Aisyme. It was later settled by people from the island of Thasos, which implies a coastal location.[239] Makessa is a rare poetic form for a Macedonian woman, but the eponymous Emathia is otherwise unknown. Neapolis (modern Kavalla) was the closest mainland point to Thasos, and was probably founded around 600 BC perhaps as a Thasian outpost.[240] The island of Thasos is about 25 km. to the southeast and is the northernmost Aegean island. It had an early Phoenician presence. The Greek settlement, including the role of the eponymous Thasos (probably the son of Kadmos) was described in detail by Herodotus, who visited the island.[241] This passage in the *Periodos* (lines 661–3) is a good example of the awkwardness that the author could not escape at times due to metrical strictures, with the word "Thasos" three times in three lines.

Lines 664–78

Thrace (Thraikia in the *Periodos*) is a general term for the regions north of the Aegean, extending in its broadest sense as far as the Istros (Danube) River. At the time that the *Periodos* was written there were localized kingdoms in the southern portions, but to the north there was a vast territory that was still little known to the Greek world. Abdera (modern Palaia Avdiron) was on the coast northwest of Thasos. It was associated with the eighth labor of Herakles, the elimination of the man-eating horses of the local ruler Diomedes, which had killed Herakles' companion Abderos. The city was allegedly founded at his grave.[242] Later it was reestablished by the Ionian city of Teos.

The Nestos River flows into the Aegean about 20 km. west of Abdera. Lake Bistonis (modern Vistonis) lies just to the east ("toward the sunrise") of Abdera, and the indigenous Bistonians, who survived into the fifth century BC, were the ethnic group of King Diomedes.[243] Maroneia (at modern Agios Charlabos) is about 25 km. east of the lake, a town of the local Kikones (who also lived in nearby Ismaros). Odysseus was active in the region and fought against them. The settlement from the island of Chios was probably in the seventh century BC.[244]

Lines 679–95

The island of Samothrace (Samothraike in the *Periodos*), about 45 km. due south of Maroneia, was discussed in unusual detail. Its mythological history began with the brothers Dardanos and Iasion, whose mother Elektra was one of the Pleiades. The Homeric account of Iasion records that he had a consensual relationship with Demeter, but Zeus nevertheless eliminated him. The version presented in the *Periodos* is somewhat less mythological, perhaps referring to theft of cultic treasures, a regular problem on the island. Dardanos, it seems, had to leave for some unspecified reason and ended up on the Trojan Plain where he founded the city of Dardania, a locality that still existed in the fifth century BC. He also established the Trojan royal line, and brought the Samothracian mysteries to Troy. Clearly there was some connection between Samothrace and the Trojan region.[245] But the account is unclear and lines 687–8 are corrupt. There was also confusion between Samothrace and the island of Samos; the word *samos* means "rugged" and was often applied to Greek islands. Nevertheless an ethnic connection between the two islands was often assumed.[246]

Lines 696–712

Ainos (modern Enez) was 50 km. east of Maroneia on the coast. It was at the Aegean end of a route to the Black Sea, and was settled at an uncertain date from the island of Mytilene.[247] The Thracian Chersonesos, or the Peninsula of Thraikia, is the long narrow feature that extends into the northeastern Aegean along the north side of the Hellespont. The description in the *Periodos* is counter-clockwise around it. Kardia was a Milesian outpost (with some help from another Ionian city, Klazomenai). The Athenians arrived in the sixth century BC as part of the elder Miltiades' local adventures. It was destroyed at the end of the fourth century BC by Lysimachos, the companion of Alexander the Great, who then founded Lysimacheia at a nearby location (probably at modern Örtaköy at the narrows of the peninsula), using part of the population of Kardia.[248]

Limnai ("Marshes") was at the outer end of the peninsula at an uncertain location.[249] Alopekonnesos was a few kilometers toward the north. Elaious ("Olives") was at modern Eski Hissarlik, inside the Hellespont, and was perhaps an Athenian outpost of the seventh century BC.[250] The name of the founder, Phorboon, is peculiar and may be corrupt. Sestos (at modern Yalıkabat) and Madytos (at modern Eçeabat) lie about seven kilometers apart on the northern coast of the Hellespont and were both founded from the island of Lesbos in the seventh century BC. They controlled the narrows of the strait.[251] Krithote ("Barley") has not been located, but was near Paktye, and both were on the southern side of the inner Chersonesos opposite Kardia and Lysimacheia, further examples of the adventures of the elder Miltiades.[252]

Lines 713–47

The next series of toponyms is along the Thracian coast of the Propontis, the sea between the Hellespont and the Thracian Bosporos. Perinthos (at modern

Ereğli) was roughly at the midpoint of the Propontis coast. Although it was allegedly founded by the island of Samos, there was an earlier settlement since it retained a pre-Greek name. Selembria (at modern Silivri, 25 km. to the east, may have been the earliest Greek town on this coast, since special note was made of the fact that it was established before Byzantion. It was another Megarian foundation. Both cities were from the seventh century BC. Ancient Byzantion was within the confines of modern Istanbul, but limited to south of the Golden Horn.

Nothing is recorded in the *Periodos* about the Bosporos, and the account next discusses the Thracian coast of the Black Sea. Unusual for Greek sea names, it was called just "The Sea" (Pontos) rather than a toponym that reflected its location: this may be because it was the largest secondary feature in the Mediterranean system. The original name was Axenos (line 735), a hellenized version of an indigenous term, *aesaena*, meaning "dark" or "sombre." But this sounded like "Inhospitable" in Greek, so it was changed to Euxeinos, or "Hospitable," yet the more generic Pontos was regularly used.[253]

Demetrios of Kallatis (see also line 117) was probably a major source for this section of the *Periodos*. Kallatis (at modern Mangalia) was on the west coast of the Black Sea. Demetrios is hardly known today: only six fragments of his 20 books on Asia and Europe survive.[254] He was active in the second century BC, so not much earlier than the date of the *Periodos*.

Philia is not certainly identified but may be the small promontory of Kara Burun on the coast of the Black Sea between the entrance of the Thracian Bosporos ("mouth of the Pontos," line 722) and Salmydessos. This is a long straight coast as well as a town (at modern Midye), which is located to the northwest. The region was infamous for shipwrecks, and the locals prospered from collecting the debris.[255]

Cape Thynias (modern İğneada Burnu) is another promontory farther along the coast, in the region known as Astike, a vague toponym or ethnym scattered throughout this district. Although the primary source remains Demetrios of Kallatis (line 719), this portion of the account seems particularly nautically oriented and may be derived from a *periplous* of the Black Sea.

Apollonia (modern Sozupol) was a Milesian foundation; the association with the reign of Kyros (Cyrus) of Persia indicates a date of around 600 BC. Such a type of synchronism has not appeared previously in the *Periodos* and may be a feature of Demetrios' technique. There are over 20 known Milesian settlements on the coast of the Black Sea.[256]

Mt. Haimos is the modern Balkan range of Bulgaria, which, although impressive, is substantially less rugged than the Tauros Mountains of Kilikia. Haimos hardly extended from the Black Sea as far as the Adriatic, but it was believed that both seas could be seen from its summits.[257] This is by no means the truth but an anecdote that established the range as particularly formidable.

Mesembria (modern Nesebur) was a few kilometers north of Apollonia across the Bay of Burgas. It was a joint settlement of Byzantion, Chalkedon, and Megara, dating to around 600 BC, on a site that was already a Thracian town.[258] The Getians (whose name may be a version of "Goths") ranged widely across Europe, from the Alps to the Black Sea as well as north of the Istros (Danube). They had first been encountered when the Greeks settled along the western Black Sea coast, and became better known when Dareios I of Persia penetrated to the Istros in the late sixth century BC.[259] The Krobyzoi were along the coast between Mesembria and the Istros.[260]

After citing the Krobyzoi, and making the suggestion that Mt. Haimos extends to the Adriatic (line 747), there is no further surviving text of the *Periodos*, since the final sheets of the manuscript have long been lost. The material that follows (lines 748–1026) is based on the reconstruction of Aubrey Diller.[261]

Lines 748–70

Archaeological evidence indicates that Odessos (modern Varna) was founded in the 560s BC, which conforms to the connection with the era of Astyages of Media, who was on the throne during much of the first half of the sixth century BC.[262] The use of another synchronism, as at lines 731–3, shows the same source, probably Demetrios. For the Krobyzoi, the indigenous population of the region, see line 746. Dionysopolis was about 35 km. farther along the coast from Odessos, near modern Balčik. It and Krounoi ("Springs") were separate towns a short distance apart, with the latter the seaport.[263] The anecdote about the statue is probably an aetiology for the establishment of a cult through the miraculous appearance of a divine image.

The Skythians were the population north of the inhabited world, and the term was often used generically for those beyond the Istros (Danube). Bizone (near modern Kavarna) lies just east of Dionysopolis, on the coast, and is little known, as demonstrated by the vagueness of the account in the *Periodos*. For Mesembria, see line 739. Kallatis (modern Mangalia) has already been noted as the home of Demetrios (line 719), the probable source for much of this material. It was established by Herakleia Pontika (on the southern coast of the Black Sea, line 1016) in the sixth century BC; another synchronism dates its founding to the reign of Amyntas of Macedonia, who came to the throne late in that century.[264]

Tomeoi (Tomeus or Tomis, modern Constantza), famous as the place of exile of Ovid, lies on the coast 40 km. beyond Kallatis, and was a Milesian settlement of the sixth century BC, perceived as being deep into Skythian territory.[265] Istros (modern Histria) is about 40 km. farther at the southern edge of the lagoons and channels that mark the outlet of the Istros River. It was the earliest Greek settlement along this coast, founded by Miletos in the mid-sixth century BC, after the Kimmerian movement south into Asia Minor, with archaeological remains from later in the century.[266] The Kimmerians were northern nomads who pushed into Asia Minor in the seventh century BC, and who impacted on the Phrygian kingdom of Midas and the Lydian kingdom of Gyges, also remembered as the biblical people of Gomer.[267]

Lines 771–800

The Istros River, at 2,850 km. second only to the Volga among the longest rivers of Europe, originates in the Black Forest region of Germany and flows into the Black Sea along the modern Bulgarian–Romanian border. The lower portion had long been known to the Greek world but the upper parts, north of the Alps, called the Danuvius, were only vaguely understood at the time of the writing of the *Periodos*: the name "Danuvius" was first cited by Julius Caesar.[268] The reported mouths of the river varied from two to seven. The author of the *Periodos*, recording five, was following Herodotus and Ephorus.[269] The differing numbers are due to the methods of counting and changing topography.

That a branch of the river reached the Adriatic was a persistent belief, due to the closeness to one another of affluents of both river and sea (lines 191–5), a point of view popularized by Apollonios of Rhodes in the third century BC, when he reported that the Argonauts returned to the Mediterranean by such a route.[270] The comments on the flow of the river are difficult to understand, perhaps due not only to the author's summarization of earlier material from Demetrios (line 793), but the even later synthesis of the *Periodos* by the author of the *Periplous of the Euxine Sea*. The ultimate origin of the material is Herodotus' discussion of the volume of major rivers, where he reported that the Istros was the same height in both summer and winter (making an obvious contrast with the Nile).[271] Herodotus' argument is complex, noting that the increase due to summer rainfall was counteracted by solar evaporation (more intense in the hot summer), which meant that the river was thus no higher in summer than in winter. Implicit in the discussion is the failure of people from the Mediteranean to comprehend a world where the rainy season was in summer, not winter.

The island of Peuke ("Pine") is by no means the size of Rhodes, yet the term indicates a Greek presence, although it was not cited by name until the third century BC. The complexity of the delta region of the river, which is about 100 km. across and has many channels and islands, may have led to the assumption that its islands were unusually large. Mention of Rhodes by name may be due to an inadequate summary of material from Apollonios, who was associated with the Mediteranean island and described Peuke.[272]

The island of Achilles, also known as Leuke from the white cliffs that rise suddenly out of the sea, is modern Zmeinij, east of the delta of the Istros. According to the epic known as the *Aithiopis*, Achilles' mother Thetis took him from his funeral pyre and buried him there. As an isolated island it would have attracted birds and would also have been sacred.[273] The distance of 400 stadia (about 80 km.) is far greater than the 40 km. of today, reflecting the siltation of the Istros outflow since antiquity.

Diller was unable to make a meaningful restoration of lines 796–8, and the references to the Thracians cannot easily be fitted into this part of the narrative, which is topographically well beyond the normal limits of the Thracian territory. The Bastarnians lived on the lower Istros, extending north to the upper Vistula.[274] When the text becomes coherent again the narrative is at the Tyras River (modern Dniester), whose mouth is about 90 km. beyond the north edge of the Istros delta. It originates in western Ukraine and flows for 1,362 km. southeast to the Black Sea. The source was known as early as the fifth century BC, which shows that the upstream trading presence had been established by that time. The river runs through a broad and fertile flood plain, passing the city of Tyras (at modern Bilhorod-Dnistrovskyy) located slightly inland at a headland on the right bank. The town had been established by Milesians perhaps around 600 BC.[275]

Lines 801–31

The Borysthenes River (modern Dnieper) was known to Greeks since the fifth century BC and is the second largest stream emptying into the Black Sea. Its source, 2,201 km. from its mouth, is near modern Smolensk, and traders went upstream to the point that weather conditions were believed to make further travel impossible. This river was also known for its fertile and prosperous floodplain, and was the southern part of a route to the Baltic. It emptied into a large estuary which included, to the west, the outlet of the Hypanis River (modern Bug). This region of the two rivers and their estuaries attracted extensive Greek settlement, most notably the town of Olbia (modern Parutyne), actually on the east side of the Hypanis estuary near where it joins that of the Borysthenes. From the Black Sea ships would sail up the common estuary and then turn left and go up the Hypanis estuary; the 240 stadia (about 50 km.) would be the total distance from the sea to Olbia. The names Olbia and Borysthenes seem to have been virtually interchangeable. It was the farthest north Greek settlement, founded by Miletos in the sixth century BC and becoming a great trading center where at least seven languages were spoken.[276] The site has been extensively excavated.

The Racecourse of Achilles is a sandy spit on the south side of the Borysthenes/Hypanis estuary, today about 2 km. wide or less and 65 km. long, but different in antiquity. Achilles exercised on the feature, and there was a shrine to him at the end.[277] This was the region of the Taurians, who lived in the mountainous interior of the modern Crimea (the Tauric Peninsula of the *Periodos*). They were familiar to the Greek world as the protectors of Iphigeneia after she was miraculously rescued by Artemis at Aulis, events best known from Euripides' two Iphigeneia plays. The local divinity came to be identified with Artemis or Iphigeneia. Despite their essentially pastoral character the Taurians harassed coastal cities well into the Roman period and were infamous for brutally sacrificing to the goddess Greek sailors who had been shipwrecked or captured, placing their heads on poles above their houses.[278]

The unnamed Hellenic city is probably Chersonesos (near modern Sevastopol), founded, perhaps in the fifth century BC, from Herakleia Pontika and the island of Delos, which had trade interests in the northern part of the inhabited world.[279] For Herakleia and the Kyaneans, see lines 1016–19.

Lines 832–41

Theodosia (modern Feodosia) is at the southeastern corner of the Chersonesos (Crimea). It was settled from Miletos early in the sixth century BC. The Bosporanian kingdom ("Bosporos" in the *Periodos*) was a Greek state that controlled territory on the north shore of the Black Sea and around the Maiotis (modern Sea of Azov). It came into being in the fifth century BC when the local Greek cities joined together to present a united front against the Skythians and others to their north, and effectively lasted into late antiquity.[280]

Kimmerikon (possibly at modern Opuk) is on the coast about 130 km. northeast of Theodosia and was a Greek settlement of about 500 BC, presumably at a site already occupied by the local Kimmerians. Pantikapaion (modern Kerch) was probably also established at an indigenous town. It lay on the west side of the strait known as the Kimmerian Bosporos, the passageway out of the Black Sea into the Maiotis. Its important location attracted Greek traders as early as the seventh century BC, and in 480 BC it became the capital city of the Bosporanian kingdom. It was here that Mithridates VI the Great died in 63 BC, bringing his sprawling Pontic kingdom to an end.[281]

Greek traders penetrated north from the towns on the Black Sea, as much as 40 days up the Borysthenes (lines 805–7), but it was known that a large region lay beyond any point that the Greeks had reached. This was established by the third century BC when Eratosthenes determined the size of the earth and its inhabited portions. In fact, Olbia, the northernmost Greek settlement, was only slightly more than halfway from the equator to the north pole. Thus the vastness of the north was realized, if not truly comprehended, and was generically called Skythike, after Ephorus' classification of the Skythians as the barbarians of the north.[282]

Lines 842–59

The *Periodos* catalogues populations in the interior lands north and west of the Black Sea. The primary source for this material was Ephorus.[283] There are also certain parallels to Herodotus' account of the same region, but differences, which suggest that there may be unnamed writers in the transmission from

Herodotus through Ephorus to the *Periodos*. The Karpides probably lived just north of the Istros, somewhat in the interior; they may be the Kallipidai of Herodotus, the first ethnym on his list, although he placed them on the Borysthenes, not the Istros. The Neuroi were beyond them and by implication the northernmost population.

Hybla may be the Hylaia ("Woodlands") of Herodotus, a descriptive term for the territory east ("toward the sunrise") of the Borysthenes.[284] The Georgoi ("Farmers") were allegedly beyond the Woodlands. After a deserted region there were cannibals ("Anthrophagian Skythians") and then more deserted territory. The alternation of deserted and populated areas probably does not reflect an overall view of the landscape, but the linear nature of trade routes, going from settlement to settlement. This territory was so little known that the indigenous toponyms and ethnyms were often not recorded and descriptive Greek names were used, if any at all.

The Pantikapes River is probably one of the northern affluents of the Borysthenes, perhaps the modern Inhulets, which provided a direct route to the interior. Yet the suggestion is that the traders crossed it well above its mouth, eventually reaching the Marsh People ("Limnaioi"), a topographically appropriate name for the locals, who were probably in central Ukraine.

Belief in pastoralism and a milk-oriented diet, as well as the communal nature of the locals, had been part of the Greek perception of these populations since the time of Homer. Milk was a limited part of the diet of the Greco-Roman world, with mare's milk even rarer, and was mostly found in rural situations. Thus a culture that relied on it was a curiosity.[285] Anacharsis was a Skythian aristocrat of the sixth century BC who came to the Greek world, resulting in extensive cultural interchange. Although it is difficult to separate the real personality from a vast amount of semi-legendary material that grew up around him, he nevertheless was an important figure in early Greek understanding of the remote north.[286]

Lines 860–74

There are some textual problems at this point, but it is clear that the topic is the movement of various Skythian populations. The Sakai, whose name is linguistically similar to the Skythai (Skythians), came into Asia Minor and Mesopotamia in the seventh century BC. The Sauromatai, known to the Greek world since the fifth century BC, lived beyond Lake Maiotis; their territory extended 15 days to the north, and there is a hint that their homeland went beyond the northern tree line. The Gelones were probably also in the Maiotis region, as were the Agathyrsoi. The eponyms of these people were the brothers Skythes, Gelonos, and Agathyrsos, sons of Herakles and the snake woman Echidna.[287]

Lake Maiotis (the modern Sea of Azov) is the northern extension of the Black Sea, reached through the Kimmerian Bosporos. The indigenous

Maiotai were fishing and farming people, who moved seasonally between the coast and the interior.[288] The Tanais River (modern Don) flows into the Maiotis at its northeast end. The Araxis River is unlikely to be the Araxes (modern Aras or Arax) that flows from Armenia toward the Caspian Sea, despite the assertions of previous commentators, unless the topography is exceedingly confused. The Araxis of the *Periodos* is probably an unidentified major affluent of the Tanais, or even the Rha (modern Volga), which comes within a few kilometers of the Tanais at modern Volvograd.[289]

Hekataios of Teos (one of the Ionian cities) is presumably the scholar from Abdera (a foundation of Teos), who was active in the fourth century BC and wrote *On the Hyperboreans* about these northern regions.[290] The Tanais originates at modern Novomoskovsk, 1,160 km. above its mouth. It is unlikely that the source was known to Greek traders, who penetrated into a marshy region that may have appeared to be a lake and the beginning of the river. Today there are several recognized mouths, all just west of modern Rostov-on-Don.

Lines 875–94

Further comments on the Tanais reveal that it was considered to be the boundary between Europe and Asia, something established by the fifth century BC, although the Phasis River (modern Rioni in Georgia) was a less-common alternative. Using Demetrios of Kallatis and Ephorus as his primary sources, the author of the *Periodos* mentioned two ethnic groups that lived along the Tanais.[291] The first was the Sarmatai, or Sauromatai (for which, see line 864), who held the lower 2,000 stadia (about 400 km.), perhaps as far upstream as the vicinity of modern Volvograd, where the river bends sharply to the northwest. The implication is that the Iazamatai were farther upriver, but associating them with the Maiotis also suggests a possible location along its eastern shore. These populations were nomadic and only vaguely known to the Greek world, and could have been encountered in widely separated places at different times.

The semi-mythical Amazons have exercised a powerful influence on cultural history from ancient to modern times: they had been known to the Greek world since the period of Homer. They were a mixture of history and myth, warrior women who were seen to fulfill roles normally thought to belong to men. Like all semi-mythical populations, they lived at the edge of—or just beyond—the known world, and their homeland tended to move into more remote areas as civilization advanced. They were famous for their horsemanship, and were actually believed to have been the first to practice it, which would tend to place them among the horse nomads and societies with a wealthy female elite that existed north of the Black Sea. But they were also said to have penetrated Asia Minor, and fought a major battle with Herakles—part of his ninth labor—at the Thermodon River (modern Terme Çay) on the north shore of the Black Sea.[292] *Gynaikokratoumenoi*—here an ethnym—is from Greek political terminology (*gynaikokratia*, or "rule by women"), a concept originally meant to describe states such as Sparta where it was believed that women had excessive power to a pernicious effect. The use of the term ethnically is rare; if Ephorus actually used it, he was probably the first.[293]

Leaving the Maiotis region, the account mentions two cities on the Asian side of the Kimmerian Bosporos. Hermonassa (at modern Taman) is about midway through the strait, on the south shore of a bay. It had been founded by the mid-sixth century BC but its origins are obscure.[294] Phanagoreia lies about 30 km. to the east on the same bay, and was founded slightly later. It was the most important city east of the Kimmerian Bosporos, and there are extensive remains. The *Periodos* reported that it was one of the few settlements from the island of Teos, perhaps the only foundation of that city in the Black Sea region: Hekataios of Teos or Abdera (line 870) may have been the source for this information. Sindikos Harbor, where the city of Gorgippia was, is actually outside the Bosporos, about 65 km. to the southeast of its Black Sea outlet, on the coast at modern Anapa. It was a wealthy trading center for the hinterland. The topography of the *Periodos* is rather loose at this point, for the implication is that Sindikos Harbor is actually along the east side of the Bosporos, but the entire district east of the strait and for some distance along the coast was called Sindike, and thus the author may have considered Sindikos Harbor to be on the Bosporos, a region discussed through line 900.[295] Much of the Maiotis shoreline, especially on the eastern side, is a maze of islands and channels.

Lines 895–913

There is probably a change of sources at line 895, since the city of Kimmeris is actually the Kimmerikon mentioned at line 874, located on the European side of the mouth of the Kimmerian Bosporos. Yet the toponym, referring to the indigenous local Kimmerians, may have been used (in varying forms) for several settlements in the region. Strabo seemed to know of a Kimmerikon near Phanagoreia.[296] The association with the rulers of the Bosporos indicates a foundation after the mid-fifth century BC when the Bosporanian dynasty was established at Pantikapaion.

Kepos ("Garden"), often recorded in the plural (Kepoi), was a wealthy Milesian settlement slightly northeast of Phanagoreia, perhaps founded in the mid-sixth century BC.[297] As noted, the Sindians were an indigenous population east of the Bosporos, here presented as a sub-group of the Maiotai.

The account leaves the Bosporanian region and moves southeast along the coast of the Black Sea. The Kerketai were presumably the first ethnic group encountered in this direction.[298] But the most important indigenous population, lying beyond (southeast) of the Kerketai, were the Achaians, placed along an extensive stretch of coastline at the western end of the Caucasus. Whatever their actual name was, it sounded like the Greek ethnic group mentioned prominently by Homer, as well as the toponym and ethnym eventually localized on the northern coast of the Peloponnesos. This generated a rich mythology attempting to explain how the Achaians on the Black Sea were really Greek in origin, something for which there presumably was no physical evidence. The arguments were purely linguistic, since the people were said to have lost any Greek characteristics but had become barbarized (for which the author of the *Periodos* used the rare word *ekbarbaromenous*).

The most common tale regarding the placement of Greeks on the northeastern Black Sea coast was connected to Jason and the Argonauts, since some Thessalians on his expedition were said to have settled there. But it was also believed that after the Trojan War the contingent from Orchomenos in Central Greece, led by Ialmenos—mentioned by Homer as a minor player in the war—and consisting of Minyans, ended up on this coast due to adverse winds. This story was first documented by Pherekydes in the fifth century BC. Passing over the question of how someone sailing from Troy to Central Greece could have ended up on the far Black Sea coast, the tale—which, unlike many other homecoming accounts in the epic tradition, seems to lack context—is almost certainly post-Homeric and evolved only after Greek settlers in the region encountered the population with their Greek-sounding name. Nevertheless the Achaians, whatever their origin, were known for their hostile nature even into the Roman period, when they repeatedly caused difficulty for Mithridates VI of Pontos.[299]

Lines 914–49

Farther along the coast were the Heniochoi, whose name also appears to be Greek ("Charioteers"), but again was probably nothing more than a hellenization of an indigenous ethnym. Nevertheless, as with the Achaioi, a complex mythology was used to explain their presence, with a further association with Jason and the Argonauts. The Heniochoi were said to be descended from the charioteers of the Dioskouroi (Polydeukes and Kastor), who were on the expedition. Their names vary in the sources: this is the earliest extant citation of Amphitos and Telchis, and none of the known names is likely to represent an early tradition.[300] Their presence was reinforced by the toponym Dioskourias (not mentioned in the extant *Periodos*), at modern Sukhumi, an indigenous town that had a Greek presence from at least the fifth century BC and which was a major emporium where 70 or more ethnic groups came from throughout the Caucasus region and beyond to trade.[301]

There is a brief allusion to the Kaspian (Caspian) Sea, which, at its closest, is about 650 km. from the Black Sea. It had become known to the Greek world with the last expedition of Cyrus the Great of Persia in 522 BC.[302] Its shores had been explored by the time the *Periodos* was written, but there was little if any Greek settlement on it, and thus it was outside the author's area of interest, other than the sparse and unexplained reference to the Horse Eaters, perhaps a vague comment on the horse-oriented cultures of the Black Sea region.

Presumably the stated Median boundary was somewhere north of the Phasis River on the coast of the Black Sea. The author of the *Periodos* may have meant the boundary of the Median satrapy of the Persian empire, although this was along the Kaspian Sea, and the satrapy on the Black Sea coast was actually the Armenian.[303] There was a repeated tendency in Greek and Roman literature to cite the Medes when the Persians were actually meant.

The Phasis River (modern Rioni) enters the Black Sea at its southeast corner, and was early known to Greek traders, who saw it as a major route to the east.[304] It originates on the southern slopes of the Caucasus. Iberia is the territory south of the mountains and inland from the upper Phasis. The movement of its peoples south into Armenia was one of many such changes in locations of populations occuring at various times: Strabo provided a lengthy list.[305]

The city of Phasis lay on the left bank of the homonymous river, a few kilometers inland. It was founded by Milesians, perhaps as early as the late sixth century BC. Phasis was a major trading center that drew upon the hinterland to its east, allegedly receiving goods from as far as Baktria and Indike (India). Whether this actually happened—and the evidence is disputed—it is clear that Phasis was the Mediterranean contact point for a vast network that may have extended at least into central Asia, with many ethnic groups reaching the city.[306]

Several additional ethnic groups appear in the *Periodos* after citation of the Phasis river and city. These are along the Black Sea coast as it makes it turn from southerly to westerly, but presented in loose geographical order. As has been normal, local ethnyms may have been turned into Greek descriptive names. Some cannot easily be identified. The inhabitants of Koraxike (perhaps "Raven Land") may have exported wool to the Aegean.[307] The Kolike and the Melanchlainoi (perhaps "Black Robed People") remain obscure. The best known are the Kolchoi, or Colchians, famously associated with Jason and Medea and living along the Phasis River. The Makrones were improbably said to have come from the Aegean island of Makris, or Euboia, or the name may be descriptive of their stature.[308]

Along the southern coast of the Black Sea were the Mosynoikoi ("Tower Dwellers," a hybrid Persian-Greek ethnym). They had been encountered by Xenophon in 400 BC, who left a detailed discussion of his interaction with them as well as their habits and unusual architecture. But the peculiar tale of how the king was shut up in his tower and not given food may simply be an embellishment, perhaps by Ephorus, of Xenophon's account of his attack on the Mosynoikoi citadel, which resulted in the king's failure to leave and being burned to death.[309]

Lines 950–81

The account continues west along the south shore of the Black Sea. Pharnakia (at modern Giresun) was founded by Pharnakes I of Pontos, probably in the 180s BC. The text is deficient at lines 951–2, and the mention of a "deserted place" is enigmatic, since it seems that the town was located at the existing site of Kerasous.[310] The Island of Ares is modern Giresun Adası, lying just offshore and one of the few islands on this coast, which would have helped to provide a sheltered anchorage for Pharnakia.

The Tibarenians lived in the mountains above Pharnakia and had been part of the Persian empire. The comment about their approach to life is taken almost verbatim from Ephorus.[311] Amisos (at modern Samsun) was about 180 km. farther west along the coast from Pharnakia, and was established by Miletos at the northern end of a trade route from interior Asia Minor. It was the most important city on this portion of the coast. There is no other evidence for Phokaian participation in the settlement, but if the information is accurate, this would have been the only known Phokaian outpost on the Black Sea. The *Periodos* date for the establishment of the city is based on that of Herakleia (see lines 1016–19), which was at the time of the accession of Cyrus of Persia in 559 BC, although it may have been somewhat earlier than the four years suggested.[312]

The Leukosyroi (seemingly "White Syrians") were a feature of this region; a Syrian (i.e. Assyrian) presence on this coast had been noted since at least the fifth century BC. This is well after the collapse of the Assyrian empire at the end of the seventh century BC, and those on the Black Sea coast may have been remnants who survived nearly two centuries later. There was also a town named Assyria east of Amisos. Thus it seems obvious that Assyrian traders had reached the Black Sea, but the epithet *leukos*, documented as early as Pindar, cannot be explained.[313]

The author of the *Periodos* was correct in identifying the narrowest part of Asia (Minor) as extending from the vicinity of Amisos to the Issic Gulf (the modern Gulf of Alexandretta), the northeastern corner of the Mediterranean, a distance of about 775 km. Alexandroupolis is more commonly Alexandria by Issos (modern Iskenderun) on the Issic Gulf, the first major coastal town encountered if one were to go south from Amisos. The author is the only one to attribute its founding to Alexander the Great (if this is who is meant by the "Macedonian"); there is surprisingly little information about the origins of the city, and it may actually have been a creation of Antigonos I Monopthalmos or Seleukos I, in the generation after Alexander.[314] The comment that Herodotus did not know about the "isthmus" of Asia Minor is rather irrelevant: he merely wrote that there was a trade route from Kilikia (the easternmost territory on the Asia Minor shore of the Mediterranean) north to Sinope on the Black Sea west of Amisos.[315] This was an ancient way to access the Black Sea from the Levantine regions that had probably existed from Assyrian times if not before.

107

The summary of the ethnic groups of Asia Minor is a close rendition of a list provided by Ephorus that may have passed through Apollodoros of Athens. Strabo also produced a similar account.[316] His catalogue and that of the *Periodos* are remarkably similar: both begin with the three ethnic divisions of the Greeks (Ionian, Dorian, and Aiolian), and then have the indigenous populations (12 in the *Periodos*, 13 in Strabo). Ten appear on both lists: unique to Strabo are the Bithynians, Trojans, and Milyans, and the Kappadokians and Lydians are only in the *Periodos*. Unlike Strabo, the author of the *Periodos* did not update his list to account for the arrival of the Galatians in the third century BC.

The list in the *Periodos* seems at first to follow a geographical order, beginning with the Kilikians at the southeastern corner of Asia Minor (referring back to line 968), and then continues west past the Lykians and Karians to the Aegean. But next are the Mariandynians in the northwest corner of Asia Minor, skipping over the populations on the Aegean, including Bithynia, the home of the patron of the *Periodos*. It is probable that the author did not want to include this important territory, so central to the production of the treatise, in a historic list of ethnyms. After the Mariandynians, the list follows no obvious geographical order, and was perhaps structured for metrical reasons: Strabo's list, and thus in all probability Ephorus' original, is more geographically coherent. Nevertheless, the catalogue shows the remarkable ethnic diversity of Asia Minor.

Lines 982–97

The Halys River (modern Kızıl İrmak) is the longest river in Asia Minor (1,355 km.), flowing in an arc through the central part of the territory and emptying into the Black Sea between Amisos and Sinope. It was considered the boundary between the Mediterranean world and that of the east.[317] The preserved name is Greek and descriptive ("Salty"); the indigenous name is unknown. To the west of its mouth was the city of Sinope (modern Sinop), at the northernmost point of Asia Minor, long established at the end of a trade route from the south. The *Periodos* has an unusually detailed historical ethnology of the city, which allows reconstruction of the process of its foundation over several generations. It was believed to have had an Amazonian connection, but this only became popular in the era of Alexander the Great and probably has no early origin.[318] Autolykos and his companions were associated with Herakles and his local fight with the Amazons, and became the mythical founders of the city. Many years later there was a Milesian attempt at settlement under Abron, which was unsuccessful because of the instability caused by the Kimmerian incursions of the mid-seventh century BC. A later expedition, with assistance from the island of Kos, and those Milesians who had survived the Kimmerians, eventually established the permanent settlement shortly after 600 BC.[319]

Lines 998–1011

Karambis (modern Kerembe Burnu), about 145 km. west of Sinope, was an important navigational point for crossing the Black Sea to the Bosporanian region on its north side. This was the closest point to the north shore at Krioumetopon ("Ram's Forehead," modern Cape Sarych), the southernmost point of the Chersonesos (modern Crimea). The distance across is about 275 km., and the sailing time of 24 hours would be at about 6 knots, a high speed but not impossible.[320]

About 75 km. farther along the coast from Karambis was Amastris. Phineus, the son of Phoinix of Tyre, was the mythical dynast of this region, and eventually the Milesians arrived and established several towns along the coast, not named in the *Periodos* but probably Tios, Kytoron, Kromna, and Sesamous, all located in a stretch of 80 km. Amastris was the niece of Dareios III, the last king of Persia. One of her several marriages was to Dionysios of Herakleia (see line 1016), who died in 306/305 BC. Eventually she left the city and founded Amastris, bringing the four Milesian towns together into a new city at or near Sesamous. A remarkable personality of her era, she was a rare example of a woman as city founder.[321]

Lines 1012–26

The Parthenios River (modern Bartin Su) is a small stream whose mouth is just west of Amastris. The author of the *Periodos* may have mentioned it because it was the eastern boundary of Bithynia, the territory of his patron. The Baths of Artemis, presumably pools or springs, have not been located.

Herakleia, usually called Herakleia Pontika to distinguish it from the numerous other cities dedicated to the hero, was the most important town on the western part of the Asia Minor coast of the Black Sea, located at modern Ereğli, about 50 km. southwest of the mouth of the Parthenios River. Details of its foundation are hardly known, but most sources attribute it solely to the Megarians.[322] Ephorus, however, reported Boiotian participation, but no Boiotian town was named, and it is probable that some scattered settlers from nearby Boiotia joined the Megarian expedition. The synchronism with the accession of Cyrus of Persia provides a date of around 559 BC. The Kyaneai are the modern Örektaşi islets on the east side of the Black Sea entrance to the Thracian Bosporos. They are about 175 km. from Herakleia, and mention of them seems peculiar (when several places between those points are in the following lines of the *Periodos*), but their citation along with "setting forth from Hellas" reveals the vestiges of a sailing route from the Aegean through the Bosporos, and then turning past the islands to reach Herakleia.

The Hypios River (modern Melen Çayı) is another small stream, whose mouth is about 40 km. southwest of Herakleia. Citing the river allowed the author to refer to the city of Prousias, at modern Üsküb in the hills about

15 km. east of the upper Hypios. It was founded by Prousias I of Bithynia in the early second century BC.[323] By mentioning this relatively unimportant town, the author was drawing attention to the dynastic history of Bithynia— Prousias was the great-grandfather of his patron Nikomedes III—and presumably positioning himself, as with the previous citation of the Parthenios River and the following mention of the Thynoi, for the discussion of Bithynia (now lost) that would soon follow in the *Periodos*.

The Sangarios (modern Sakarya) is the most important river in northwest Asia Minor. Its source is in Phrygia, 824 km. above its mouth. The Thynoi and their territory of Thynis would have been on the upper river near the eastern border of Bithynia; the names are related, and again the author was drawing his readers' attention to the forthcoming examination of his patron's territory.

The island of Apollonia is probably modern Kefken Adası, just off the coast about 100 km. west of Herakleia. The city of Thynias is otherwise unknown. With these various citations of toponyms and ethnyms related to the Thynoi, the reconstructed extant *Periodos* comes to an end, probably less than halfway through its full extent. Nevertheless the author has made it clear that a detailed discussion of Bithynia was the next major topic.

Notes

1   Marcotte, *Géographes*; M. Korenjak, *Die Welt-Rundreise eines anonymen griechischen Autors ("Pseudo-Skymnos")* (Hildesheim 2003).
2   Diller, *Tradition* 113.
3   *EANS* 999–1002.
4   *SIG* 585.86; F. Gisinger, "Skymnos [1]," *RE* 5 (second series 1927) 661–72.
5   August Meineke, *Scymni Chii periegesis et Dionysii descriptio Graeciae* (Berlin 1846) 42; F. X. Ryan, "Der sogennante Pseudo-Skymnos," *QUCC* 87 (2007) 137–43.
6   Constantine, *On the Themes* 1.2. p. 17; *FGrHist* #854, T1; Aubrey Diller, "The Authors Named Pausanias," *TAPA* 86 (1955) 276–9.
7   *FGrHist* #244; Marcotte, *Géographes* 44–6.
8   *FGrHist* #396; K. Boshnakov, *Pseudo-Skymnos* (Stuttgart 2004) 33–69; Georgia L. Irby-Massie, "Semos of Delos," *EANS* 730–1; Graham Shipley, "Three Studies of 'Pseudo-Skymnos'," *CR* 57 (2007) 349.
9   *BNP Chronologies of the Ancient World* 99–100.
10 Livy 44.14.5–7.
11 Polybios 36.15; Strabo, *Geography* 14.1.38; Justin, *Epitome* 34.4; Appian, *Mithridates* 4–7.
12 *OGIS* 345; David Magie, *Roman Rule in Asia Minor* (Princeton 1950) 318; Richard D. Sullivan, *Near Eastern Royalty and Rome, 100–30 BC* (Toronto 1990) 30–3.
13 Marcotte, *Géographes* 7–16; Jessica Lightfoot, "'Not Enduring the Wanderings of Odysseus': Poetry, Prose, and Patronage in Pseudo-Scymnus' *Periodos to Nicomedes*," *TAPA* 150 (2020) 406–7.
14 Sullivan, *Near Eastern Royalty* 33–5.
15 Polybios 36.15; Justin, *Epitome* 34.4; Appian, *Mithridateios* 4–7.
16 Marcotte, *Géographes* 24–35.

17 Dionysios of Halikarnassos, *On Literary Composition* 25.
18 Lightfoot, "Not Enduring" 381–9.
19 Diogenes Laertios 9.20.
20 Lionel Pearson, *Early Ionian Historians* (Oxford 1939) 16.
21 Valery Yailenko, "Source Study Analysis of Pseudo-Scymnus' Data on The Pontic Cities' Foundation," *Pontica* 18–19 (2015–16) 9–23.
22 Marcotte, *Géographes*; Diller, *Tradition* 165–76.
23 Diller, *Tradition* 165.
24 Richard Hunter, "The Prologue of the *Periodos to Nicomedes* ('Pseudo-Scymnus')," *HG* 11 (2006) 123–4; Lightfoot, "Not Enduring" 389–94.
25 Getzel M. Cohen, *The Hellenistic Settlements in Europe, the Islands, and Asia Minor* (Berkeley 1995) 400–2.
26 Empedokles, F1 Graham; Lucretius 1.62–71.
27 *FGrHist* #244, F25, 53.
28 P. M. Fraser, *Ptolemaic Alexandria* (Oxford 1972) vol. 1, p. 471.
29 Daryn Lehoux, "Diogenes of Babylon," *EANS* 253.
30 Athenaios 15.671f.
31 Aristotle, *Rhetoric* 2.23.11; Suetonius, *Grammarians* 10.
32 Polybios 5.33.5.
33 Lionel Casson, *Libraries in the Ancient World* (New Haven 2001) 49–52.
34 Eratosthenes (*FGrHist* #241) F1a; Marcotte, *Géographes* 9–11.
35 Eratosthenes, *Geography* F30–1.
36 K. Tuchelt, "Didyma," *PECS* 272–3; H. W. Parke, *The Oracles of Apollo in Asia Minor* (London 1985) 55.
37 Polybios 12.26d.2.
38 Polybios 34.1.5.
39 Katherine Clarke, *Between Geography and History: Hellenistic Constructions of the Roman World* (Oxford 1999) 120.
40 Herodotus 4.42; Roller, *Ancient Geography* 50–1.
41 Strabo, *Geography* 1.1.1.
42 Aristotle, *Meteorologika* 2.5. etc.
43 Homer, *Odyssey* 1.3; Lightfoot, "Not Enduring" 394–5.
44 Roller, *Eratosthenes' Geography* 13–14.
45 *FGrHist* #70; Roller, *Ancient Geography* 80–3; Serena Bianchetti, "Aspetti di geografia ephorea nei *Giambi a Nicomede*," *PP* 69 (2014) 751–80.
46 Dionysios (*FGrHist* #1773) T1.
47 *FGrHist* #85.
48 Paul T. Keyser, "Kleon of Surakousai," *EANS* 481.
49 Duane W. Roller, "Timosthenes of Rhodes," in *New Directions in the Study of Ancient Geography* (ed. Duane W. Roller, University Park 2019) 56–79.
50 *FGrHist* #124; Waldemar Heckel, *Who's Who in the Age of Alexander the Great* (Oxford 2006) 76–7.
51 *FGrHist* #566.
52 Clarke, *Between* 95.
53 Diller, *Ancient Measurements* 6–9.
54 Poseidonios F246; Roller, *Historical and Topographical Guide* 168–9.
55 Avienus 350–5; Henry Mendell, "Euktemon of Athens," *EANS* 317.
56 Strabo, *Geography* 3.5.5.
57 Strabo, *Geography* 3.4.2; Roller, *Historical and Topographical Guide* 151.
58 Marcotte, *Géographes* 158–60; María Eugenia Aubet, "Mainake: The Legend and the New Archaeological Evidence," in *Mediterranean Urbanization 800–600 BC* (ed. Robin Osborne and Barry Cunliffe, Oxford 2006) 187–202.

59   Hesiod, *Theogony* 290; Herodotus 4.8; Ephorus F129a; Strabo, *Geography* 3.5.4.
60   Ephorus F128; supra, p. 17.
61   Velleius 1.2.3; Pomponius Mela 3.46; Carolina López-Ruiz, "Tarshish and Tartessos Revisited: Textual Problems and Historical Implications," in *Colonial Encounters in Ancient Iberia* (ed. Michael Dietler and Carolina López-Ruiz, Chicago 2009) 263–6.
62   Strabo, *Geography* 3.2.7, 15.2.12–13.
63   Pliny, *Natural History* 9.11–13.
64   Herodotus 4.152; Pausanias 6.19.2; López-Ruiz, "Tarshish" 255–80.
65   Strabo, *Geography* 2.5.15; Roller, *Historical and Topographical Guide* 110.
66   Ephorus F30 = Strabo, *Geography* 1.2.28; Kosmas 2.148; Clarke, *Between* 199–200.
67   Agathemeros 2.
68   Roller, *Ancient Geography* 73.
69   Herodotus 4.49; Ephorus F131b.
70   Marcotte, *Géographes* 164–5.
71   Strabo, *Geography* 3.5.6.
72   Strabo, *Geography* 4.4.1, 5.1.4.
73   Apollonios 4.323–8; Roller, *Ancient Geography* 161–3.
74   Dietler, "Colonial Encounters" 5–8.
75   Avienus 481–4.
76   Timaios F71; Herodotus 1.163; Strabo, *Geography* 4.1.4–5.
77   Strabo, *Geography* 3.4.8.
78   Polybios 4.47.1; Strabo, *Geography* 14.2.10; Roller, *Historical and Topographical Guide* 156.
79   Guy Barruol and Michel Py, "Recherces récentes sur la ville antique d'Espeyran à Saint-Gilles-du-Gard," *RAN* 11 (1978) 19–100.
80   Boardman, *Greeks Overseas* 217–19.
81   As in Aristotle, *Meteorologika* 1.13.351a.
82   Homer, *Iliad* 2.840–1; *Odyssey* 19.177; Ephorus F113; Strabo, *Geography* 5.2.3–4.
83   Herodotus 1.94; see also Dionysios of Halikarnassos, *Roman Antiquities* 1.27; Strabo, *Geography* 5.2.1; Guy Bradley, *Ancient Umbria* (Oxford 2000) 98–9.
84   Timaios F65 = Strabo, *Geography* 14.2.10.
85   Theophrastos, *Research on Plants* 5.8.3.
86   Hesiod, *Theogony* 1011–13.
87   Hekataios of Miletos F62; Dionysios of Halikarnassos, *Roman Antiquities* 1.35.3; Strabo, *Geography* 2.5.20.
88   Polybios 1.2.7.
89   Dionysios of Halikarnassos, *Roman Antiquities* 1.72.
90   Antiochos of Syracuse (*FGrHist* #555) F7; Polybios 34.11.5–7; Strabo, *Geography* 5.4.3.
91   Boardman, *Greeks Overseas* 168–9.
92   Hesiod, *Works and Days* 635–40; T. J. Dunbabin, *The Western Greeks* (Oxford 1948) 6–7.
93   Marcotte, *Géographes* 176.
94   Strabo, *Geography* 5.4.7; Boardman, *Greeks Overseas* 192.
95   Strabo, *Geography* 5.4.11.
96   Strabo, *Geography* 6.1.1–3.
97   Dionysios of Halikarnassos, *Roman Antiquities* 1.11; Strabo, *Geography* 6.1.4.
98   Strabo, *Geography* 5.4.13; Boardman, *Greeks Overseas* 180–2.
99   Herodotus 1.167; Strabo, *Geography* 6.1.1.

100 Homer, *Odyssey* 10.1–55; Strabo, *Geography* 6.2.10–11; Pausanias 10.11.4.
101 Boardman, *Greeks Overseas* 188–9.
102 Thucydides 6.2.2; Ephorus F136; Polybios 23.13.12; Strabo, *Geography* 6.2.4.
103 Homer, *Odyssey* 11.107 etc.; Strabo, *Geography* 6.2.1.
104 Dionysios of Halikarnassos, *Roman Antiquitities* 1.22.
105 Thucydides 6.3; Ephorus F137a; Strabo, *Geography* 6.2.2.
106 Dunbabin, *Western Greeks* 8.
107 See the chart, *BNP* 3 (2003) 565–70.
108 Thucydides 6.3; Strabo, *Geography* 6.2.2; Boardman, *Greeks Overseas* 174–7.
109 Boardman, *Greeks Overseas* 172, 184–5.
110 Thucydides 6.3.2; Plutarch, *Love Stories* 2.773b; Strabo, *Geography* 6.2.4.
111 Strabo, *Geography* 6.2.6.
112 Strabo, *Geography* 6.2.3; Roller, *Historical and Topographical Guide* 308.
113 Boardman, *Greeks Overseas* 170–1.
114 Herodotus 7.154; Strabo, *Geography* 6.2.6.
115 Boardman, *Greeks Overseas* 186–7.
116 Thucydides 6.4.4; Strabo, *Geography* 6.2.5; Boardman, *Greeks Overseas* 187–8.
117 Marcotte, *Géographie* 185.
118 Salvatore De Vincenzo, "Sicily," in *The Oxford Handbook of the Phoenician and Punic Mediterranean* (ed. Carolina López-Ruiz and Brian R. Doak, Oxford 2019) 537–52.
119 Herodotus 1.145 etc.
120 Timaios F13a; Polybios 2.39; Valerius Maximus 8.7.ext. 2; Gianfranco Maddoli, "The Concept of 'Magna Graecia' and the Pythagoreans," in *Brill's Companion to Ancient Geography* (ed. Serena Bianchetti *et al.*, Leiden 2016) 44–5.
121 Strabo, *Geography* 6.1.5; Giuseppina Spadea, "Terina e lo Pseudo-Scimno," *PP* 29 (1974) 81–3.
122 Thucydides 5.5.3; Strabo, *Geography* 6.1.5.
123 Strabo, *Geography* 6.1.6; Roller, *Historical and Topographical Guide* 294–5.
124 Aristotle, *Politics* 2.9.5; Ephorus F139; Diodoros 12.20; Strabo, *Geography* 6.1.8.
125 Strabo, *Geography* 9.3.1; Roller, *Historical and Topographical Guide* 539.
126 Hekataios of Miletos F84; Strabo, *Geography* 6.1.10; Boardman, *Greeks Overseas* 180.
127 Ovid, *Metamorphoses* 15.9–59; Strabo, *Geography* 6.1.12; Diogenes Laertios 8.3.
128 Diodoros 12.6.9–10.
129 Cicero, *de finibus* 5.4; Pausanias 6.19.11; Boardman, *Greeks Overseas* 180.
130 Strabo, *Geography* 6.3.1–3; Barclay V. Head, *Historia Numorum* (Oxford 1911) 54–5; W. D. E. Coulson, "Taras," *PECS* 878–80.
131 Diodoros 8.18; Varro, *de re rustica* 1.44.2; Athenaios 12.519–20; Boardman, *Greeks Overseas* 178–9.
132 Hekataios of Miletos F91–2; Strabo, *Geography* 7.5.9.
133 Hekataios of Miletos F86–8; Herodotus 4.99; Thucydides 7.33.4; Strabo, *Geography* 6.3.1; Pausanias 10.10.6.
134 Strabo, *Geography* 6.3.6.
135 Theopompos (*FGrHist* #115) F132; see also Pseudo-Skylax 16; Bradley, *Ancient Umbria* 21–2.
136 Theopompos F130.
137 Herodotus 4.33.
138 Carolynn E. Roncaglia, *Northern Italy in the Roman World* (Baltimore 2018) 9–10.
139 *On Marvellous Things Heard* 128.
140 Aristotle, *Meteorologika* 1.9.

141 Homer, *Iliad* 2.852; Livy 1.1.2; Vergil, *Aeneid* 1.242–53; Strabo, *Geography* 5.1.4.
142 Hesiod, *Theogony* 338; Herodotus 3.115; Ovid, *Metamorphoses* 2.1–366; Strabo, *Geography* 5.1.9.
143 Polybios 2.16.13; Plutarch, *On the Slowness of Divine Vengeance* 12.
144 J. F. Healy, *Mining and Metallurgy in the Greek and Roman World* (London 1978) 60–1.
145 Pseudo-Skylax 21–2; Stephanos of Byzantion, "Mentores."
146 Strabo, *Geography* 7.5.4, 7.7.8.
147 Timaios F77; Eratosthenes, *Geography* F146; Avienus 152–4.
148 Diodoros 15.14; Strabo, *Geography* 7.5.5.
149 Polybios 2.2–12; Marcotte, *Géographes* 203.
150 Diodoros 15.13.4–5.
151 Apollonios 4.569–71; Marcotte, *Géographes* 204.
152 Juba of Mauretania (*FGrHist* #275) F5, 60; Strabo, *Geography* 6.3.9, 7.7.7.
153 Herodotus 6.45.
154 Thucydides 1.24–6.
155 Herodotus 5.61; Euripides, *Bakchai* 1330–9; Apollodoros, *Bibliotheke* 3.5.4; Strabo, *Geography* 7.7.8.
156 Boardman, *Greeks Overseas* 226.
157 Homer, *Odyssey* 14.314–20 etc.
158 Thucydides 2.80; Strabo, *Geography* 7.7.5.
159 Plutarch, *Greek Questions* 11; Boardman, *Greeks Overseas* 225–6.
160 Plutarch, *Pyrrhos* 1; Heckel, *Who's Who* 181–3.
161 Homer, *Iliad* 2.750 etc.; Strabo, *Geography* 7.7.10–12.
162 Marcotte, *Géographes* 208.
163 Strabo, *Geography* 7.7.7; Apollodoros, *Bibliotheke* 3.7.7.
164 Thucydides 2.102; Strabo, *Geography* 10.2.2, 6.
165 Herodotus 8.45; Thucydides 1.30; Strabo, *Geography* 10.2.8.
166 Homer, *Odyssey* 20.210 etc.
167 Strabo, *Geography* 10.2.10.
168 Strabo, *Geography* 10.2.18.
169 Herodotus 2.10; Strabo, *Geography* 10.2.19.
170 Ephorus F122a, F144; Strabo, *Geography* 10.3.1–4.
171 Thucydides 2.91–2; N. G. L. Hammond, "The Peloponnese," *CAH* 3.1 (1982) 696–744.
172 Strabo, *Geography* 8.3.33.
173 Strabo, *Geography* 9.4.8.
174 Pausanias 2.4.3.
175 Strabo, *Geography* 9.2.5.
176 Arrian, *Anabasis* 1.7–9; Plutarch, *Alexander* 11–12.
177 Ovid, *Metamorphoses* 10.605; Pausanias 1.39.5.
178 Homer, *Odyssey* 3.286–90; Strabo, *Geography* 8.6.20.
179 Strabo, *Geography* 8.5.1; Pausanias 3.25.4–8.
180 Polybios 2.16.3, 7.6.1; Strabo, *Geography* 2.1.35; Roller, *Ancient Geography* 76.
181 Strabo, *Geography* 8.8.5; Marcotte, *Géographes* 214.
182 Strabo, *Geography* 8.6.19; Pausanias 2.3.2.
183 Ephorus F18b.
184 Diodoros 7.9.2; Strabo, *Geography* 8.8.5.
185 Polybios 2.41.4–5.
186 Pausanias 5.4.1–4, 5.3.6.
187 Pausanias 4.3.
188 Strabo, *Geography* 8.5.4.

189 Nikolaos of Damascus (*FGrHist* #90) F30; Pausanias 2.12.6, 2.28.3.
190 Ephorus F145; Strabo, *Geography* 10.4.1.
191 Homer, *Iliad* 2.649; *Odyssey* 19.174.
192 Homer, *Odyssey* 19.172–202; Richard Hunter, "Pseudo-Scymnus," in *Hellenistic Poetry: A Selection* (ed. David Sider, Ann Arbor 2016) 534.
193 Homer, *Odyssey* 19.175–7; Strabo, *Geography* 10.4.6.
194 Casson, *Ships* 281–96.
195 Homer, *Odyssey* 9.81.
196 Pindar, *Nemean* 4.46; Herodotus 8.49; Strabo, *Geography* 8.6.16; Apollodoros, *Bibliotheke* 3.12.6; Roller, *Historical and Topographical Guide* 474–5.
197 Herodotus 8.44.
198 Aristophanes, *Acharnanians* 75.
199 Johannes Scherf, "Cecrops," *BNP* 3 (2003) 59–60.
200 Ephorus F151; Dionysios of Chalkis (*FGrHist* #1773) F9.
201 Herodotus 1.171; Strabo, *Geography* 13.1.59.
202 Strabo, *Geography* 10.1.8; Plutarch, *Greek Questions* 22.
203 Homer, *Iliad* 2.538.
204 Strabo, *Geography* 8.6.13; Pausanias 4.34.9–11.
205 Homer, *Iliad* 2.539; Herodotus 4.33; Strabo, *Geography* 10.1.6.
206 Homer, *Iliad* 2.537.
207 Diodoros 5.79.2; Strabo, *Geography* 9.5.16; Roller, *Historical and Topographical Guide* 571–2.
208 Diodoros 5.74.
209 Strabo, *Geography* 9.4.2; Apollodoros, *Bibliotheke* 1.7.2.
210 Hesiod, *Catalogue of Women* F9; Diodoros 5.80.2; Strabo, *Geography* 9.4.10–11.
211 Thucydides 3.92; Strabo, *Geography* 9.4.13.
212 Herodotus 7.200; Strabo, *Geography* 9.3.7.
213 Hellanikos of Lesbos (*FGrHist* #4) F51; Euripides, *Phoenician Women* 666–75, 818–21; Strabo, *Geography* 9.5.10; Apollodoros, *Bibliotheke* 3.4.1.
214 Homer, *Iliad* 2.683.
215 Homer, *Iliad* 2.757–8.
216 Homer, *Iliad* 2.752–4.
217 Diodoros 4.18.6–7.
218 Diodoros 14.82.7; Strabo, *Geography* 9.4.11.
219 Strabo, *Geography* 9.5.5.
220 Strabo, *Geography* 9.5.23.
221 Hesiod, *Catalogue of Women* F7.
222 Strabo, *Geography* 7.7.4, 8.
223 Hesiod, *Theogony* 339.
224 Marcotte, *Géographes* 223.
225 Herodotus 7.123; Xenophon, *Hellenika* 5.2.13; Strabo, *Geography* 7.F11.
226 Livy 45.30.5; Strabo, *Geography* 7.F14; Acts 17:10–13.
227 Cohen, *Hellenistic Settlements in Europe* 101–5.
228 Thucydides 1.137; Diodoros 13.49.1–2; Strabo, *Geography* 7.F11.
229 Cohen, *Hellenistic Settlements in Europe* 95–9.
230 Diodoros 16.53–5; Cohen, *Hellenistic Settlements in Europe* 91–2.
231 Strabo, *Geography* 7.F16.
232 Herodotus 7.123; Ephorus F34.
233 Herodotus 7.122.
234 Homer, *Iliad* 1.590–4, 14.230; Herodotus 6.140; Apollodoros, *Bibliotheke* 1.9.17.
235 Thucydides 4.84.1; B. S. J. Isserlin *et al.*, "The Canal of Xerxes: Summary of Investigations 1991–2001," *BSA* 98 (2003) 369–85.

236 Thucydides 4.102, 5.6.1; Diodoros 12.32.3.
237 Diodoros 31.8.8.
238 Eratosthenes, *Geography* F13; Strabo, *Geography* 1.3.1, 2.3.5, 7.F16; Plutarch, *How One Might Become Aware of His Progress in Virtus* 7.79a.
239 Homer, Iliad 8.304; Thucydides 4.107; Diodoros 12.68.
240 Pseudo-Skylax 67.1; Strabo, *Geography* 7.F16; Boardman, *Greeks Overseas* 232.
241 Herodotus 6.46–7; Marcotte, *Géographes* 227.
242 Strabo, *Geography* 7.F18; Apollodoros, *Bibliotheke* 2.5.8.
243 Herodotus 7.110.
244 Homer, *Odyssey* 9.39–61; Boardman, *Greeks Overseas* 230.
245 Homer, *Iliad* 20.215–40; *Odyssey* 5.125–8; Hesiod, *Catalogue of Women* F121; Herodotus 5.117; Dionysios of Halikarnassos, *Roman Antiquities* 1.61; Strabo, *Geography* 7.F20.
246 Strabo, *Geography* 10.2.17; Pausanias 7.4.2.
247 Boardman, *Greeks Overseas* 230.
248 Herodotus 6.36; Diodoros 20.29; Strabo, *Geography* 7.F21; Cohen, *Hellenistic Settlements in Europe* 82–7.
249 Strabo, *Geography* 14.1.6.
250 Boardman, *Greeks Overseas* 265.
251 Boardman, *Greeks Overseas* 241.
252 Strabo, *Geography* 7.F21; Herodotus 6.35–40.
253 Stephanie West, "'The Most Marvellous of All Seas': The Greek Encounter With the Euxine," *G&R* 50 (2003) 151–67.
254 *FGrHist* #85.
255 Xenophon, *Anabasis* 7.5.12–14.
256 Boardman, *Greeks Overseas* 225–6; a list of Milesian settlements appears in *BNJ* 3 (2003) 567–8.
257 Theopompos (*FGrHist* #115) F129; Polybios 24.4.
258 Strabo, *Geography* 7.6.1; Boardman, *Greeks Overseas* 247.
259 Herodotus 4.93.
260 Hekataios of Miletos F170; Herodotus 4.49.
261 Diller, *Tradition* 165–76.
262 Boardman, *Greeks Overseas* 247.
263 Pomponius Mela 2.22.
264 *BNP, Names, Dates, and Dynasties* 188.
265 Boardman, *Greeks Overseas* 247.
266 Boardman, *Greeks Overseas* 247–9; Marcotte, *Géographes* 241–2.
267 Herodotus 1.6, 15; Strabo, *Geography* 1.3.21; Ezekiel 38:6.
268 Hesiod, *Theogony* 339; Caesar, *Gallic War* 6.25.
269 Herodotus 4.47; Ephorus F157; Eratosthenes, *Geography* F148; Strabo, *Geography* 7.3.15.
270 Apollonios 4.288–93.
271 Herodotus 4.50.
272 Apollonios 4.309–12; Eratosthenes, *Geography* F148–9.
273 *Aithiopis, Argument* 4; Euripides, *Iphigeneia Among the Taurians* 438.
274 Strabo, *Geography* 2.1.41; Pliny, *Natural History* 4.81.
275 Herodotus 4.51; Boardman, *Greeks Overseas* 250.
276 Herodotus 4.24; Strabo, Geography 7.3.17; Boardman, *Greeks Overseas* 242–3, 250–1.
277 Strabo, *Geography* 7.3.19; Pliny, *Natural History* 4.83; Ptolemy, *Geographical Guide* 3.5.7.
278 Herodotus 4.103.

279 Herodotus 4.33–5; Boardman, *Greeks Overseas* 251–2.
280 Boardman, *Greeks Overseas* 252–3; Iris von Bredow, "Regnum Bosporanum," *BNP* 12 (2008) 443–50.
281 Diodoros 12.31.1; Strabo, *Geography* 7.4.4; Boardman, *Greeks Overseas* 253; Duane W. Roller, *Empire of the Black Sea* (Oxford 2020) 211–12.
282 Ephorus F30a.
283 Herodotus 4.17–19; Ephorus F158.
284 Marcotte, *Géographes* 139.
285 Homer, *Iliad* 13.5–6; Andrew Dalby, *Food in the Ancient World From A to Z* (London 2003) 217–18.
286 Herodotus 4.76–8; Diogenes Laertios 1.101–5; J. F. Kindstrand, *Anacharsis: The Legend and the Apophthegmata* (Uppsala 1981).
287 Herodotus 1.103–6; 4.9–10, 21.
288 Herodotus 4.123; Strabo, *Geography* 11.2.4.
289 Strabo, *Geography* 11.4.2; Marcotte, *Géographes* 250–1; J. D. P. Bolton, *Aristeas of Proconnesus* (Oxford 1962) 59–60.
290 Hekataios of Abdera (*FGrHist* #264) F10, 13.
291 Herodotus 4.45; *Airs, Waters, and Places* 13; Ephorus F160; Demetrios of Kallatis (*FGrHist* #85) F1.
292 Homer, *Iliad* 3.189, 6.186; Herodotus 4.110–17; Strabo, *Geography* 11.5.1–4; Adrienne Mayor, *The Amazons* (Princeton 2014) 170–2; Anna A. Trofimova, ed., *Greeks on the Black Sea* (Los Angeles 2007).
293 Ephorus F160a; Aristotle, *Politics* 2.6.6, 5.9.6; Strabo, *Geography* 3.4.18; see also Pseudo-Skylax 70–1.
294 Boardman, *Greeks Overseas* 254.
295 Strabo, *Geography* 11.2.10; Roller, *Historical and Topographical Guide* 636–7.
296 Strabo, *Geography* 11.2.5.
297 Boardman, *Greeks Overseas* 254.
298 Strabo, *Geography* 11.2.14.
299 Homer, *Iliad* 2.511–12, 9.82; Pherekydes (*FGrHist* #3) F143; Strabo, *Geography* 9.3.42, 11.2.12; Appian, *Mithridateios* 67, 102.
300 Strabo, *Geography* 11.2.12.
301 Strabo, *Geography* 11.2.16.
302 Hekataios of Miletos F291; Herodotus 1.203.
303 Herodotus 3.92–3.
304 Hesiod, *Theogony* 340.
305 Strabo, *Geography* 1.3.21.
306 Boardman, *Greeks Overseas* 254; David Braund, *Georgia in Antiquity* (Oxford 1994) 40–1.
307 Hipponax F2.
308 Scholia to Apollonios 2.392; Braund, *Georgia* 36.
309 Ephorus F161b; Xenophon, *Anabasis* 5.4.1–26.
310 Arrian, *Periplous of the Euxine Sea* 16.4; Cohen, *Hellenistic Settlements in Europe* 387–8.
311 Herodotus 3.94, 7.78; Ephorus F43.
312 Boardman, *Greeks Overseas* 255.
313 Pindar F173; Herodotus 1.72; Pseudo-Skylax 89.1; Apollonios 2.964.
314 Getzel M. Cohen, *The Hellenistic Settlements in Syria, the Red Sea Basin, and North Africa* (Berkeley 2006) 73–6.
315 Herodotus 2.34.
316 Ephorus F162; Apollodoros F170; Strabo, *Geography* 14.5.23.
317 Aeschylus, *Persians* 865; Herodotus 1.6, 75; Strabo, *Geography* 12.3.12.

318 Herodotus 2.34; Andron of Teos (*FGrHist* #802) F3; Marcotte, *Géographes* 259–60.
319 Strabo, *Geography* 12.3.11; Plutarch, *Lucullus* 23.4–6; Boardman, *Greeks Overseas* 254–5.
320 Casson, *Ships* 281–96.
321 Strabo, *Geography* 12.3.9; Cohen, *Hellenistic Settlements in Europe* 383–4; Heckel, *Who's Who* 21.
322 Xenophon, *Anabasis* 6.2.1; Ephorus F44a; Arrian, *Periplous of the Euxine Sea* 13; Stanley M. Burstein, *Outpost of Hellenism: The Emergence of Heraclea on the Black Sea* (Berkeley 1976) 12–19.
323 Cohen, *Hellenistic Settlements in Europe* 406–7.

4

THE *ORA MARITIMA* OF AVIENUS

One of the more unusual texts surviving from antiquity is the fragmentary *Ora Maritima* ("The Sea Coast") of a certain Postumius Rufus Festus Avienus, written in the mid-fourth century AD. The extant 713 lines are in Latin iambic senarii, the meter of drama and comedy. Unlike the author of the King Nikomedes *Periodos*, Avienus felt no need to justify his use of poetry.

There is no manuscript tradition for the *Ora Maritima*; the existing text is that published by Victor Pisanus in 1488, containing the known major works of Avienus and some additional astronomical material.[1] The poem is a *periplous* from the region of Brittany and adjacent areas, including perhaps Ireland, and extending along the coasts of the outer edge of Europe and into the Mediterranean. The extant text ends at Massalia (Avienus' Massilia, modern Marseille), but Avienus indicated that the treatise continued as far as the Euxine (Black) Sea (lines 69–71).[2]

Other than the name of the author and that of the dedicatee, a certain Probus (lines 24, 51, 632), both of whom can be identified, there is no other internal evidence as to the date of the poem. In fact were it not for these names, it would be possible to believe that the poem dated from the first century AD, since the cited sources (lines 42–50, 117–19) are Hellenistic or earlier, and the latest personality mentioned is Juba II of Mauretania, who died in AD 23/4 (line 280). Avienus explicitly stated that his material was drawn from "old pages" (*vestutis paginis*, line 9), yet he never recorded what intermediate material may have lain between his 11 mentioned authors and his own text.

Rufus Festus Avienus

Postumius Rufus (or Rufius) Festus Avienus was a notable personality of the fourth century AD. He was of Etruscan background, from Volsinii (modern Orvieto), as he noted on his own epitaph, dedicated to the local divinity Nortia. He also recorded that he had held two proconsulships. One of these was in Achaia and the other in Africa, both of uncertain date but probably

DOI: 10.4324/9781003030379-5

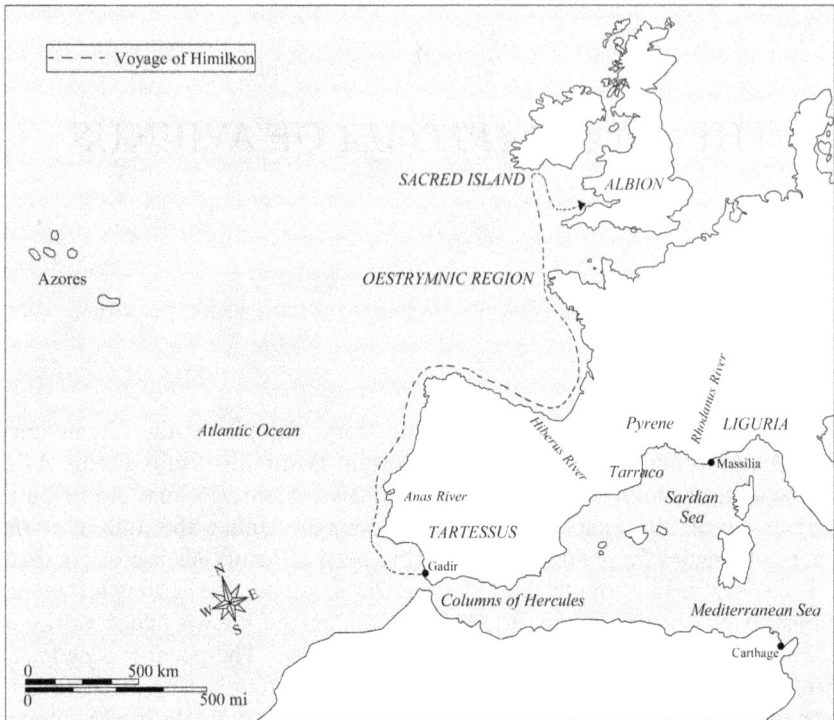

Map 4.1 Major places cited by Avienus and the voyage of Himilkon.
Source: Map by E. Rodriguez

around the middle of the fourth century AD.[3] This agrees with the career of
the dedicatee, whether it was Sextus Petronius Probus, the consul of AD 371,
or his son, Amicius Petronius Probus, the consul of AD 406. Either is possible, but the son is more likely. Avienus would have first associated with him
when he was a young man at the start of his career (lines 14–15).[4]

Avienus' literary career was largely devoted to creating Latin poetic versions
of Greek geographical and astonomical texts, as published by Pisanus in 1488.
His *Descriptio Orbis Terrae* consists of 1493 hexameters that are a paraphrase of
the *Periegesis* of Dionysios Periegetes, a didactic poem describing the entire
known world and written during the reign of Hadrian (AD 117–38). It was one
of the several texts preserved in the corpus of the *Minor Greek Geographers*,
which also included Hanno and the King Nikomedes *Periodos*.[5] Avienus' Latin
adaptation was written before the *Ora Maritima*, since there is a reference to the
former in the latter (lines 71–3).

His third major work was a Latin edition and expansion of the *Phainomena* of Aratos of Soloi, in 1878 hexameters, nearly twice as long as the original. Aratos' treatise was astronomical with close attention to weather

phenomena, and survives as one of the major extant didactic poems of the Hellenistic era, written in the early third century BC.[6] Influential in its own right, it had long attracted the interest of Romans, and Latin versions also exist by Cicero and Germanicus (the son of Drusus and Antonia).

Geography and the Latin language

As a Greek invention, the discipline of geography created a Greek vocabulary for its technical terminology, first established by Eratosthenes in the second half of the third century BC, and refined by his successors Hipparchos, Polybios, and Strabo. When the Romans began to investigate Greek culture, they often struggled to adapt Greek modes of expression to the Latin language. One way to solve this issue was to continue to rely on Greek scholars such as Polybios, who was commissioned by the Romans to explore coastal west Africa in 146 BC, and perhaps the routes across Europe, and published his results in Greek even if for a Roman audience.[7] But Romans also wrote in Greek, such as Juba II with his geographical ethnologies *Libyka* and *On Arabia*, or even the future emperor Claudius with his histories.[8]

Yet Latin authors had to contend with creating terminology for Greek technical terms and concepts. Transliteration was often the best solution, but there were issues of applying Latin case endings and details of orthography to Greek words. Pliny the Elder, in the mid-first century AD, would mix Greek and Latin forms in a single word, or even leave a Greek term in Greek characters.[9] Handling toponyms and ethnyms had always been a problem, even in Greek, where they were derived from one or more previous languages, often radically different in orthography: turning these into Latin meant an additional layer of confusion, so that the Persian king familiarly known as Darius came from Greek Dareios, which in turn was Old Persian Darayavaush, with other forms in other regional languages.

Such issues may have been part of the reason that Cicero never wrote his planned geography, even though he was familiar with Eratosthenes.[10] Julius Caesar knew both Eratosthenes and Hipparchos, and was astute geographically but never seems to have written a treatise specifically on the topic, although he did begin work on a map of the world, which was continued by Marcus Agrippa and completed by Augustus.[11] Geographical writing in Rome was a practical tool of the state administration, not an intellectual exercise.[12]

Although the Romans acquired and digested geographical information from at least the second century BC, the earliest extant Latin treatise on the topic did not appear until the mid-first century AD, the three-book *Chorographia* of Pomponius Mela. Even though an important work in the history of Roman geographical scholarship, it is remarkable for its brevity, in no way comparable to Greek geographical treatises such as the 17 books of Strabo's *Geography* or the 11 of the *Geographoumena* of Artemidoros of

Ephesos, written around 100 BC.[13] Moreover, the title of Mela's work, *Chorographia*, shows that Roman geographical writing tended more toward location of places than scholarly analysis, reflecting the political and economic basis of knowing the demographic and topographic extent of the new Roman world. Much the same can be said for the five geographical books of Pliny the Elder's *Natural History*, written in the AD 70s. Not even a free-standing work, these chapters are in the chorographic tradition and again replicate the brevity of Mela's work, which was a major source for Pliny.

It is perhaps no surprise, then, that for the Romans an important method of understanding Greek geographical scholarship was to create Latin adaptations. As early as the mid-first century BC, they were producing their own versions of Greek scholarly works, such as Cicero's *Aratea*, which was his edition of Aratos' *Phainomena*, followed by that ascribed to Germanicus from perhaps a generation or two later.[14] These, and other such works, provided the prototype for Latin editions of Greek treatises, a pattern that in time led to the adaptations of Avienus some centuries later.

The *Ora Maritima*

The surviving text, as printed by Pisanus in 1488, consists of 713 hexameters that describe the coast from Brittany to Massalia (Massilia), the remnant of a treatise that went as least as far as the Black Sea. It is based, to some extent, on a lost Hellenistic prototype, but unlike Avienus' other two adaptations, this cannot easily be identified since there is no extant work from which the *Ora Maritima* could have been derived. If the reference to Juba II (lines 280–3) was from the prototype, it was no earlier than the end of the first century BC, but the somewhat intrusive reference to the king seems to represent another source, and Avienus' autoptic comment at this point (line 274) demonstrates that it was almost certainly his own addition.

A catalogue of 11 sources at lines 42–50 is baffling. The names vary from the well-known (Herodotus, Thucydides) to the less common (Scylax, Damastus) and the exceedingly obscure (Cleon of Syracuse, Bacoris of Rhodes). These may have been the authorities used by the unidentified prototype of the *Ora Maritima*, but, other than the obvious, their dates are speculative, and of the lesser-known ones, only Damastes and Euctemon of Athens are cited elsewhere in the extant treatise (lines 337, 350). He can be dated to the second half of the fifth century BC, but Thucydides is somewhat later, and Bacoris of Rhodes may have been his contemporary.[15] Nevertheless, if these 11 names represent the sources of the prototype, it suggests a date of no earlier than the fourth century BC for that treatise.[16]

Several other personal names appear in the *Ora Maritima* but not in the list at lines 42–50.[17] Citation of Sallust (C. Sallustius Crispus), the political and literary personality of the mid-first century BC (line 33) is in the dedication to Probus, and is thus Avienus' addition. Plautus is cited at line 347, probably also Avienus' insertion, although this is not certain. Mention of Juba II shows that the latest datable addition to the treatise, other than the dedication to Probus, was at the beginning of the first century AD, but is probably a comment by Avienus himself. Himilco of Carthage, who went north of the Pillars of Herakles at about the same time that Hanno went south, was cited three times (lines 117, 383, 412). The implication is that there was some sort of written report of his cruise (line 383: *Himilco tradit*), but presumably this was accessed derivatively by Avienus, probably through another source than the one that provided the 11 names at lines 42–50. The *Ora Maritima* is the primary source for Himilco: the only other reference is by Pliny the Elder, who also mentioned a published report.[18] No extant Greek treatise refers to the explorer, and whatever record he left has vanished. The unidentified Epicureans cited at line 652 add little to understanding of the date of the source material.

Thus at least three sources lie behind the *Ora Maritima*. The primary one is the Hellenistic *periplous* or geographical treatise that cited the 11 predecessors, dating to no earlier than the fourth century BC. It presumably represents a Massalian perspective, given the emphasis on the Atlantic regions north of the Pillars, and was probably based on Massalian information of perhaps the sixth century BC. Then there is whatever report existed of Himilco's cruise, which took place around 500 BC but may have been summarized in a Greek version within the century. It is unlikely that this report survived long, and it does not seem to have been available to the author of the *periplous* that is the primary source for the *Ora Maritima*, but only to Pliny and Avienus. And finally there are some additions from a later date, including the references to Plautus, Sallust, and Juba, which are oddly intrusive and presumably Avienus' own commentary. In addition there is one case where he reported what he saw, at Gades (line 274).[19]

There is one other addition to the text that probably does not come from any of the 11 sources, unless one or more of them is later than suspected. This is the reference at lines 399–403 to the Caspian Sea being an inlet of the External Ocean, a significant topographical error in ancient comprehension of the world that only began at the time of Alexander the Great. This misconception lasted until medieval times, and thus any such reference to the Caspian must date after the late fourth century BC, and thus probably later than the 11 sources for the *Periplous*. It seems unlikely that this relatively minor point would be Avienus' addition, and thus it may reflect yet another source that he used, perhaps one on the seas and oceans.

Massalian exploration

The establishment of Massalia around 600 BC, at a location just east of the Rhone delta, opened up the western Mediterranean and the Atlantic north of the Pillars to Greek trade and exploration. Previous Greek activities in those regions had been sporadic, such as the voyage of Kolaios of Samos to Tartessos in 630 BC.[20] There were also reconnaissances from the Ionian city of Phokaia, whose ships and sailors went throughout the northwestern quadrant of the Mediterranean and perhaps beyond; one of these expeditions had founded Massalia.[21] The Phokaians, and then the Massalians, centered their efforts north of the Pillars since the Carthaginians had established a presence to the south but were hardly active to the north: the cruise of Himilco had little if any settlement component. But Massalia eventually became the most important Greek city in the western Mediterranean. Much of the material in the *Ora Maritima* reflects these early Phokaian and Massalian voyages of the sixth and fifth centuries BC, going as far as the land of the Hierni and the Albiones (lines 110–12). The Phokaians probably only explored along the central Iberian coast, five days beyond the Pillars (line 147), to the vicinity of Cape Roca near Lisbon, the westernmost point of Europe.[22] Yet the Massalians went farther, to the Oestrymnic Bay (line 95), probably the western extremity of Brittany. Much of the knowledge of these voyages is based on Massalian reports handed down over the centuries and eventually ending up in the *Ora Maritima*, which is the primary extant source for this material. It is perhaps no accident that the preserved portion of the treatise ends at Massalia: the remainder of the Mediterranean had been discussed in exhaustive detail by the time Avienus wrote, but as structured the extant *periplous* allows a final focus on Massalia itself, which owed its reputation to "the diligent work of its early founders" (lines 710–11). The elimination the rest of the treatise, whether by accident or design, meant that it was turned into a Massalian *periplous*, a tribute to those early Greeks who went out of the Mediterranean, perhaps as far as Ireland.

The structure of Avienus' poem

The *Ora Maritima* begins with the dedication to Probus (lines 1–31), followed by an exposition of Avienus' reason for writing the treatise and the 11 authors who were the primary sources (lines 32–50). A further address to Probus outlines the topics to be covered, but in terms of human constructions ("lofty cities") and the natural topography. The geographical extent of the poem is also mentioned: as far as Scythia (Skythia) and the Euxine (Black Sea) (lines 51–79).

The actual topographical account begins at line 80. After positioning the narrative at Gades (Gadir or Gadeira) and the Columns of Hercules (Herakles), the access points to the Atlantic (lines 82–9), the description jumps rather awkwardly to the Oestrymnic Bay at the northwest corner of Europe (lines 90–107). A digression describes the Hierni and Albiones, with reference to Himilco's presence in this region (lines 108–29). Then the account moves south along the modern French and Iberian coasts, past many places whose locations can only be suggested (lines 129–223). Eventually it reaches southwestern Iberia, in the region of Tartessus (Tartessos) and Gades (lines 223–317), and the Columns of Hercules (lines 317–74). This is followed by a digression on the External Ocean, again with reference to Himilco (lines 375–416). When the narrative returns to Iberia, it continues up the Mediterranean coast as far as Pyrene, or the Pyrenees (lines 417–529), a region receiving a detailed discussion (lines 530–65). Beyond the mountains the coastal account continues to the mouth of the Rhodanos (Rhodanus, lines 566–630), followed by a digression on the river itself and its upper course (631–99). The extant text concludes with comments about Massalia (lines 700–13).

As can be seen, the *Ora Maritima* is not totally linear, and although it is basically in the form of a *periplous*, there are several digressions, including one as far inland as the Alps (lines 637–40), as well as a summary of all of the oceans of the world, including the Caspian (lines 399–403). But the basic *periplous* form is retained, an account from northwest Europe back to Massalia, which was an itinerary that allowed Massalian sailors to return home from the farthest part of the world. In fact, the text retains conspicuous remnants of its origin as a *periplous*, with sailing times, the winds, and the tides (lines 108–9, 117, 176–80, 384–5, 623, 670–1). Coastal features are said to "rise up" (*intumescit, attolitur*, lines 183, 545, 689). The mouths of rivers and the local inhabitants on various stretches of shore were recorded (lines 288–90, 303), as well as details of landing places (lines 319, 460–1).[23] Its literary qualities have long been demeaned: "confused and confusing," "ill arranged and often obscure," or "perilously like nonsense" are all phrases modern critics have used to describe the poem.[24] Regardless of the difficulties in understanding the poem—and there are many—literary merit (or lack of it) is secondary to the vast amount of unique information Avienus provided about Greek geography and exploration.

The text

(1) Thinking, Probus, that you have often asked to be able to understand with your mind and senses the location of the Tauric Sea—so there would be a credible assurance even for those separated from it by the farthest expanse of the earth—I have performed this labor freely, so that your desire might be clarified through this poem. Moreover, I did not think it proper that at an advanced age you would not have lying before you a perception of the shape of that region, which I had acquired through a more private reading from old pages every day of my life. For to be grudging toward another about something that involves hardly any expense to yourself is considered to be boorish and stubborn. I also add to this that you are in the position of a child to me, in terms of love and the bond of blood. Yet this would not be enough if I did not know that you have always greedily absorbed literature and the obscure pronouncements of

the ancients. You are open minded and capable in your senses, with a constant thirst in your heart for such matters, and you retain inner thoughts beyond those of others. Why would I pour out ineffectively the secret of things to someone who does not retain it? And why mutter deep matters to someone who does not follow them? Thus I have calculated many things, Probus, many things, in fulfilling your ceaseless demands. Moreover, I have believed that it would be a parental duty if my Muse were to relate what you desire more amply and lavishly, since a man who is not stingy gives what is desired, but to augment the entire gift with something new shows a generous and open heart.

(32) You have asked, if you remember, where the location of the Maeotic Sea is. I knew that Sallust has provided this, and I cannot deny that his words have been judged by everyone to be considered authoritative. Thus we have taken many things from a number of commentaries, which have been joined to his famous description in which the shape and the image of the places have been presented quite effectively and truthfully, expressing the charm of his tongue and his pen.

(42) Thus Hecataeus the Milesian will be here for you, and Hellanicus the Lesbian, as well as Phileus the Athenian, Scylax the Caryandian, and also Pausimachus, whom ancient Samos bore. Then there is Damastus, born at famous Sige, Bacoris, originating in Rhodes, and Euctemon from the Attic city of the people, Cleon the Sicilian, Herodotus the Thurian himself, and finally he who holds the great distinction in eloquence, Attic Thucydides.

(51) In fact, Probus—part of my heart—you will hear about whatever islands rise up in the open sea, that is the open sea from which (after the hollow of the gaping world from the Tartessian Strait and the Atlantic waves) our sea is thrust forward as far as a distant soil, and the curved bays and promontories with their shores thrown back on themselves, with ridges extending themselves far into the waves, lofty cities washed by the sea, and from what sources the greatest streams flow forth, as the rivers rush down to enter the swirling sea, as well as the islands that they often encircle and the harbors that safe arms widely embrace, and also how the lagoons spread out, the lakes lie, and how the high mountains raise their rocky summits and the rushing white wave touches the forests. This will be the limit of our labor, expounding on the Scythian deep, the Euxine salt sea, and whatever islands rise up in that whiteness. We have written down the remainder more fully in that volume in which we set forth the coasts and parts of the world. But so that my sweat and labor be open and familiar to you, we will begin the narration of this little work somewhat more deeply, and you will

store in your innermost feelings what has been presented, for it is supported by evidence sought and drawn from ancient authors.

(80) The globe of the earth spreads out extensively, and moreover the waves flow around the globe. But where the deep saltwater inserts itself from the Ocean, so that the waters of our sea spread far, there is the Atlantic Gulf. Here is the city of Gadir, formerly called Tartessus; here are the columns of tenacious Hercules, Abila and Calpe: one is to the left of the lands mentioned, and Abila is near Libya. They resound with the harsh north wind but hold fast in place. Here rises the head of a ridge which a more ancient era called Oestrymnis, and its high mass and rocky point is completely turned toward the warm *notos* wind.

(94) Under the summit of this promontory, the Oestrymnic Bay opens for the inhabitants. The Oestrymnides Islands themselves are scattered through it, lying spread out and rich in minerals, tin, and lead. The people here have great vigor, a proud spirit, and an effective resourcefulness. All of them are constantly concerned with commerce. They traverse the broad stormy waters and the swelling Ocean, abundant in monsters, with small woven boats. They do not know how to cover keels with pine or maple, and they do not know how to hollow out light boats from fir, as is customary. Yet it is marvellous that they always fashion vessels from skins joined together and often run through the vast salt sea on hides.

(108) Furthermore, from here it is two days by ship to the Sacred Island, as the ancients called it. It is rich in turf, lying between the waves, and the people of the Hierni live spread across it. In addition, the island of the Albiones lies nearby, and the Tartessians were accustomed to carry on commerce as far as the end of the Oestrymnides. Settlers from Carthage, also, and the common people living around the Columns of Hercules were busy in these seas. Punic Himilco reported that he himself had investigated this matter on a cruise, and maintained that it can hardly be crossed in four months. No extensive winds propel a vessel, and the sluggish moisture of the immobile sea is paralyzed. He also adds that a lot of seaweed projects out of the water, which often holds back the prow like a thicket. He says that nonetheless the depth is not great here, and the bottom of the sea does not drop far and is only covered with a little water. Here and there they always encountered wild creatures of the sea, and beasts swim unexpectedly among the slow and sluggish ships.

(129) Anyone who would dare to drive a boat north into the waves from the Oestrymnic Islands to where the air of Lycaon

becomes stiff enters the Ligurian land, devoid of inhabitants, where the fields have been empty for a long time because of a band of Celts and frequent battles. The expelled Ligurians— since chance often drives some—came in terror to the dense thickets they now possess. There are many rocks in these places, sharp cliffs, and threatening mountains that thrust up into the heavens. For a long time, these elusive peoples spent their days among the close crags, withdrawn from the waves, for they were afraid of the salt sea because of the former danger. Yet later quiet and tranquility persuaded them to come down from the heights and descend to maritime places, where security strengthened their audacity.

(146) After what we discussed above, the great bay of the broad sea opens up as far as Ophiussa. In addition, from this shore to the Inner Sea—where, as I said before, the sea introduces itself onto the land and which is called the Sardus—there extends a journey of seven days on foot. Moreover, Ophiussa spreads out as widely as you hear that the island of Pelops extends in the Graian [i.e. Greek] territory. This was first called Oestrymnis, and those inhabiting the localities and the fields were the Oes- trymnici. Later those living there fled because of numerous ser- pents who gave their name to the empty soil.

(158) Then the ridge of Venus extends into the whirling waters, and the sea barks around two islands, which are uninhabitable because of the poverty of the places. Next is the Aryium Pro- montory, which swells toward the harsh north. From there, the route to the Columns of capable Hercules is five days for ships. After this, there is a maritime island that is abundant in grass and sacred to Saturn. Yet there is such a natural force on it that if someone approaches by sailing, the island soon awakens and customarily shakes, and all the salt water springs up to a height and trembles, although the rest of the sea is as quiet as a marsh. From here the Ophiussa Promontory rises into the air, and the route from the Arvian Ridge to these places extends for two days. But the bay that opens widely from there cannot easily be navigated by a single wind, since if you come to the middle, carried along by the zephyr, the remainder demands the *notus*. Again, if someone were to seek the Tartessian shore from there by foot, this would be hardly be completed on the fourth day. If the route were extended to our sea and the harbor of Malaca, the journey is five days.

(182) Then the Cempsician Ridge rises up, and the island called Achale by its inhabitants lies beneath it. It is difficult to credit the following story, because of its miraculous nature, yet it is supported by frequent authority. They say that within the con- fines of this island the appearance of the water is in no way

similar to that elsewhere. For everywhere else there is a radiance in the waves similar to the brightness of glass, and certainly there is a bluish image in the waves as far as the depths of the bright sea. But the ancients record that the sea is always stirred up there with foul mud, and dirty sediment adheres to the waters.

(195) The Cempsi and the Sefes hold the steep hills of the Ophiussian land. Next to them, the agile Ligians [?] and the offspring of the Draganes have located their homes under the exceedingly snowy north wind. And Poetanion is an island alongside the Sefumi with an open harbor. From there, the Cempsian people adjoin the Cynetes. Then there is the Cynetician Ridge, which slopes high toward where the starlight sets, the extremity of wealthy Europe that inclines toward the salty Ocean, abundant in beasts.

(205) The Ana River flows out there, through the Cynetes, and divides the soil. The bay spreads out again and the hollow land extends toward the south. The mentioned river suddenly divides into two streams, and through the thick fluid of the previous bay—for all of this sea is rich with mud—they are driven on their slow courses. There are two island summits that rise up high. The name of the smaller is unknown and the other is called Agonis, through a tenacious custom. Then a crag stands up on the rocks, also sacred to Saturn. The sea strikes against them and seethes, and the rocky shores extend broadly. Here shaggy she goats and many billy goats wander for the inhabitants through the thicket-filled land, producing long and heavy hair for the use of the camps and for naval clothing. From here to the river mentioned it is a journey of one day, and here is the boundary of the Cynetian peoples.

(223) The Tartessian territory adjoins them and the Tartessus River washes the land. From here extends the ridge sacred to Zephyrus. Then there is the height of the citadel called Zephyris, and the high summit of the peak rises from the ridge. A great swelling lifts into the air, and fog settling upon it always causes the peak to disappear in clouds. From here the entire region is completely grassy and clouds are constantly drawn over the inhabitants, with dense air and the day somewhat thick, and frequent dew in the manner of nighttime. No gusts of wind are customarily carried in, and no ethereal breeze from above disperses the heavy breath of air. Gloomy darkness lies on the earth, and the land is exceedingly damp. If someone were to depart from the citadel of Zephyris in a boat and be drawn into the waters of our sea, he would be carried forward by the blasts of Favonius.

(241) After here there is the sacred ridge of the infernal goddess and the rich shrine, the hidden cave, and the invisible sanctuary.

Nearby is a large swamp called Erebea, and then the community of Herbi, which is believed to have stood here since ancient days. Devoured by the storm of wars, it has left only its fame and name to the land. The Hiberus River flows from these places and creates fertility with its waves. Many say that the Hiberians are named from it, not from the river that glides past the restless Vascones. But whatever people lie west of the line of this river they call the Hiberi. The part toward the east includes the Tartessians and the Cilbiceni. Afterward there is the island of Cartare, and it is reasonable enough to believe that in antiquity it was held by the Cempsi, but later, defeated in war by their neighbors, they went forth seeking different places. Next Mt. Cassius rises up. Because of it, tin was formerly called *cassiterum* in the Graian language.

(260) Next there is a prominent shrine and the distant ancient citadel of Geron, which retains its Greek name, for we have heard that Geryon was once named after it. Here the edge of the Tartessian Bay spreads over a distance, and from the mentioned river to these places it is a journey of one day in boats. The town of Gadir is here, which means "fenced place" in the Punic language. It was formerly called Tartessus. In ancient times it was a large and wealthy state, but now it is poor, now it is small, now it is destitute, now a heap of ruins. Here we saw nothing remarkable beyond the rites of Hercules. But there was such strength and such honor among them in earlier times, and, as it is believed, a proud and outstandingly powerful king of all those who happened at that time to possess the Maurusian peoples, Juba, who was most favored by the princeps Octavian, and who was always involved in the study of letters, believed that he himself was more illustrious by holding the duumvirate of that city, although separated by the intervening sea.

(283) The Tartessus River spreads through openings from the Ligustine Lake and binds an island on all sides with its gliding. It does not roll forward in a simple course, or singly cleave the underlying soil, but it brings three mouths into fields on the eastern side and washes the south of the city with four mouths. But Mt. Argentarius—called this from antiquity because of its appearance—looms over the swamp. Its sides shine because of a great amount of *stagnum*, and from a distance it discharges light into the air, especially when the fiery sun strikes down onto the high ridges. The same river also rolls down waves of heavy *stagnum*, and brings rich metals to the walls.

(298) In the midst of where a vast region of land recedes from the sea of flowing salt water live the Etmanian people. From there—as far as the sowings of the Cempsians—the Ileates have spread themselves through the fertile fields, but the Cibiceni

possess the seacoast. As we have said above, salt water in the middle separates the citadel of Geron from the prominent shrine, and a bay intrudes between the high crags. At the second ridge there is a full river, fully flowing out. Then the mountain of the Tartessians rises up, dark with forests. Next is the island of Erythia, spread across the land and once under Punic rule. At first settlers from ancient Carthage possessed it. Erythia is separated from the mainland by five stadia, with the sea flowing in between. Toward the sunset from the citadel there is an island consecrated to Venus Marina with a temple of Venus on it, a cavity within it, and an oracle.

(317) When you come to the mountain that I said was forested, there lies a receding shore with yielding sand, in which the Besilus and Cilbus Rivers press the waters. After that, to the west, the Sacred Ridge raises its proud crags. At one time Graia called this Herma. Herma, moreover, is an earthen fortification, and fortifies a lake that flows from both sides. In addition, others say that this was the route of Hercules, for Hercules is said to have made the sea subside in order to open up an easy path for his captive herds. Moreover, many authors report that Herma was once solely under Libyan rule. Dionysius is not to be rejected as a witness, reporting that Tartessus is the boundary of Libya. On the European land a promontory rises up, which I said was called Sacred by the inhabitants, and a narrow strait flows between both places. Moreover, Euctemon, who was an inhabitant of the city of Amphipolis, says that the place called Herma or the Road of Hercules has a length extending no more than 108 miles and is three miles wide.

(341) Here stand the Columns of Hercules, which, as we read, are established as the boundaries of both continents. They are, moreover, rock promontories that are equal, Abila and Calpe. Calpe is on Hispanic soil and Abila on the Maurusian. The Punic people named Abila, which in a barbarian language (that is, Latin) means "high mountain": Plautus is the source for this. Calpe, however, in Graia is the name of a type of hollow and rounded-looking pitcher.

(350) Euctemon the Athenian, also, says that they are not rocks and that peaks do not rise up on both sides. He records that two islands lie between the land of Libyan soil and the shore of Europe, and says that these are called the Columns of Hercules. He reports that they are 30 stadia apart and trembling with forests, and always have been inhospitable to sailors. He further says that there are temples and altars of Hercules on them. Visiting ships sail in, sacrifice to the god, and then quickly move away, for it is believed sacrilegious to remain on the islands. He also reports that an extremely shallow and extensive sea lies next

to them and extends around them. Loaded ships cannot safely approach these places because of the shallowness and the thick mud of the shore. But if someone by chance wishes to come and approach the shrine, it is best that he take the vessel to Luna Island and remove the weight from the ship, and thus be carried over the sea in a small boat.

(370) But in regard to the seething waves that flow between the columns, Damastus [F2] says that it is scarcely seven stadia. Scylax the Caryandian [F8] asserts that the distance flowing between the columns extends as much as the flow in the Bosporus. Beyond these columns, on the European side, there were once villages and cities held by the inhabitants of Carthage. Their custom was to weave boats with flatter bottoms, so that the broader vessel would glide better over the surface of the shallow sea. But the open sea to the west of these columns has boundless waters with the Ocean extending widely and the salt waters stretching out, as Himilco reports. No one comes to these seas, and no one has brought his hull onto these waters because there are no propelling gusts on the open sea and no heavenly breath to assist the boat. Thus, because the air clothes the boat like a cloak, a mist always hides the waters and clouds last throughout a rather thick day.

(390) That is the Ocean that stretches across the washed-around world. That is the great sea; these are the waves that encircle the shores. This is the nourisher of the inner salt water; this is the parent of our sea. From outside, it forms the curves of several bays and the power of the deep glides into our world. But we will speak to you about the four greatest parts. The first of these is the insertion into the land of the Hesperian tides and the Atlantic salt water. Then there is the Hyrcana wave—the Caspian Sea—and then the salt water of the Indians and the surface of the Persian water, and the Arabian waters under the *notos*, now warm. In old usage this was once called the Ocean, then the Hyrcana wave or the Caspian Sea. Its waves unfold in a long circuit and extend at length on a wandering shore. Yet the salt water is often so shallow where it spreads out that it scarcely conceals the underlying sands. The waves are often covered and the waters restricted by marshy ground. Strong beasts swim throughout the sea, and there is a great fear of monsters living among the waters. Punic Himilco once recorded that he himself had seen and examined them in the Ocean, something published a long time ago in the innermost Punic annals that we have reported to you. Now let the pen return to the previous topic.

(417) As I have said, opposite the Libystidic Column another rises in the European land. Here the Chrysus River enters the high waters, and four peoples live on the near and far side, for

the fierce Libyphoenicians are in this place, and there are the Massieni, the kingdom of the Cilbiceni with its fertile fields, and the rich Tartessians who extend to the Calactic Bay.

(425) Next to these, soon there is the Barbetian ridge and the Malacha River, with a homonymous city; in former times it was called Menace. At that point there is an island of the Tartessians, which is under the rule of that city and which was once consecrated by its inhabitants to Noctiluca. On that island there is a lagoon and a safe harbor, and the town of Menace is above it. Where the mentioned district draws itself back from the waves, Mt. Silurus swells up, with its high summit. There an immense crag rises and enters the deep sea. The pines that were once frequent here gave it its name in the Graian language. The coast falls back as far as the shrine of Venus and the ridge of Venus. Formerly there were frequent cities on this coast, and many Phoenicians once held these lands. Now the deserted earth is spread with inhospitable sands, and the land is deprived of crops and lies waste. From the mentioned ridge of Venus, Herma of the Libyan land—as I said previously—can be seen some distance away. Here also the shore lies empty of inhabitants today and the land is abandoned. Formerly many cities stood here and many peoples filled the places.

(449) Then the Namnatian harbor—near the Massienian town— curves in from the high seas, and the city of Massiena with its high walls rises on its innermost bay. After this the ridge of Trete rises up, and next to it is the small island of Strongyle. Bordering this island is an immense marsh whose sides spread out. From there the Theodorus River bursts forth: you should accept without surprise that in such a fierce and barbarous place there is a name in the Graian language. The Phoenicians formerly lived in these places. Then the sands of the shore spread from here, and three islands broadly encircle this shore. This once stood as the boundary of the Tartessians, and here was the city of Herna. The Gymnetian peoples possessed these lands as far as the valley of the Sicanus, which flows past it. Now it is deserted and has lacked inhabitants for a long time. The Alebus River, full of its own sound, flows away. After here is the island of Gymnesia, through the waves, which gave its ancient name to its inhabiting people. Then the Pityussae Islands come into view, and the extensive ridges of the Balearic Islands.

(472) Opposite them, the Hiberi extend their rule to the ridge of Pyrene, located broadly along the interior sea. The first of their cities that rises up is Ilerda. From there the coast spreads out in sterile sands. The city of Hemeroscopium was once inhabited here, but now the region is empty of residents and is damp because of a languid swamp. Next the city of Sicana rises up,

called this by the Hibericans from a nearby river. Not far from where this river divides, the Tyrius River touches the town of Tyris. But where the land recedes far from the salt water, the region extends into a broad overgrown interior. The Berybraces, a wild and fierce people, wander among their frequent herds of livestock. They nurture themselves hardily with milk and rich cheese, and sustain their lives in the fashion of wild animals.

(489) After this, the Crabrasia Ridge advances to a height and the bare shores lie all the way to the boundaries of Onussa Cherronesus. The marsh of the Naccarares extends through these places. Custom has given this name to the marsh, and there is a small island rising up in the middle of the swamp. It is rich in olives and thus is sacred to Minerva. There were many cities around here, since Hylactes, Hystra, Sarna, and famous Tyrichae (the ancient name of the town) stood here. The treasures of the inhabitants are well remembered throughout the regions of the world. In addition to the fertility of the earth, through which the land produces livestock and palm groves, and the gifts of golden Ceres, foreign goods are carried up the Hiberus River.

(504) Next the Sacred Mountain extrudes its proud head and the Oleum River cuts through the nearby fields, flowing between the twin peaks of ridges. In fact, Mt. Sellus—this is the ancient name for the mountain—rises up to the heights of the clouds. In an earlier era the city of Lebedontia stood next to it, but now the region is devoid of Lares and endures the haunts and dens of wild beasts. After these places, sands lie over a wide area where the town of Salauris once stood, and among them there was once ancient Callipolis, that Callipolis which, because of its extensive and high walls and tall summits, rises into the heavens, and whose great perimeter of habitation presses on a marsh rich in fish on both sides.

(519) Then there is the town of Tarraco and the pleasant sea of the rich Barcilones. For there the harbor spreads out into safe arms and the land always becomes wet with sweet water. After it, the harsh Indigetes present themselves, those hard people, a fierce people attracted to the hunt and to moving around. Then the Celebantic Ridge extends its back as far as salty Thetis. Now there are only so many rumors that the city of Cypsela once stood there, and the rough land preserves no vestiges of the former city.

(530) There the harbor yawns open with a very large bay, and the salt sea creeps broadly into the hollow land. After this, the Indicetic shore reclines as far as the summit of the Pyrene promontory. After the shore that we said lies in a low-lying region,

Mt. Malodes extrudes into the waves, with two rocks swelling up, and twin summits seek their heights in the clouds. Between them a harbor lies spread out and the sea is not exposed to any gusts. Thus, since the crags are placed far out, the summits of the peaks go all around, the waves lie immobile over the rocks, the sea is quiet, and the enclosed waters are calm.

(544) Then there is the Toni lagoon at the foot of the mountains, and the ridge of the Tononitian cliffs rises up, through which the loud Anystus River churns the foamy sea and divides the salt water with its flood. Because of the waves, this strait is also salty. Whatever lands submit to the deep waters were all formerly held by the Ceretes and the harsh Ausoceretes, but now by people with a name in common, the Hiberi. Then there are the Sordian people, who live in remote places and extend as far as the interior sea where the pine-bearing summits of Pyrene stand. They spend their days among the haunts of wild beasts and make their impression extensively in the fields and on the waters of the sea. It is reported that on the boundary of the Sordicene land once stood the city of Pyrene, with its rich houses, and that the inhabitants of Massilia often conducted business with them. But to Pyrene from the Columns of Hercules, the Atlantic waves, and the boundary of Zephyris, is a run of seven days for a fast ship.

(565) After the Pyrenaic ridge, there lie the sands of the Cynetic shore with the Rhoscynus River widely dividing them. As we have said, this is the land of the Sordicenian earth. Here there is a lagoon and a swamp that spreads out broadly, which the inhabitants call the Sordices. Beyond the noisy waters of the vast swirling waves—since because of the extensive circumference of its broad edge it often swells with driving winds—the Sordus River flows out from this marsh. Then from its mouth... . The shore is then curved by the deep sea and the land is hollowed by its own losses. The wave creeps along more generously and the great mass of the waters spreads out. Three very large islands stand in it and the sea is filled with hard rocks. Not far away from there another bay of broken earth gapes and the sea encircles four islands, which in former usage were called the Piplae. Formerly the Elesycian people held these places, and the city of Naro was the great head of a fierce kingdom. Here the Attagus River rushes down to the salt sea, and then the Helice swamp is nearby. Next, as ancient reports have recorded, stood Besara. Yet now the Heledus and Orobus Rivers—demonstrative of former beauty—pass through empty fields and heaps of ruins. Not far from these the Thyrius rolls into the sea... . The rolling sea is never disturbed and the quiet of Alcyone always spreads over the waves. But the top of this crag extends from this region

to the promontory that I said was called Candidum. Nearby is the island of Blasco, where the earth is consumed by the sea in a rounded shape. On the mainland amid the peaks of rising ridges the sandy soil extends back and the shores—deprived of inhabitants—spread out.

(608) Then high Mt. Setius swells up, pine-bearing with its citadel. The ridge of Setius with its broad base extends as far as the Taurus, and the people call the swamp near the Oranus River the Taurus. The Hiberian land and the harsh Ligurians are divided by it. Here sits the small settlement and poor town of Polygium. Then there is the village of Mansa, the town of Naustalo, and the city on the sea of the Haesician people [?]... .

(621) The Classius stream flows into that sea. But the Cimenice region recedes far from the salty stream, over a broad region thick with forests. The meaning of the name is "high mountain with woods on its back." The Rhodanus touches the lower slopes with its stream and wanders through the sharp massive projections that overhang the sea. The Ligurians extend themselves to the waves of the Internal Sea from the citadel of Setiena and the distant cliffs of the rocky ridge.

(630) But I must speak to you more about the Rhodanus River, lingering as I drag my pen along, my Probus, and reveal where the stream has its source, the gliding of the wandering waters, what peoples the waves of the river lap at, how much profit it provides for the inhabitants, as well as speaking of the divisions at its mouth.

(637) The Alps raise their snowy ridge to the heavens, toward the sunrise, and the fields of Gallic soil are divided by its rocky points, where the breathing winds are always tempestuous. It [the Rhodanus] flows from here and is driven from its source, cutting through a gaping cave with violent force. The surface of the waters is navigable at its first rising and extremity, but that side of the ridge which is elevated and produces the river the locals call the Column of the Sun. It rises up into the clouds to such a height that the southern sun is scarcely visible due to the barrier of the ridge when it approaches the northern region to return the day. You know that this was the opinion of the Epicurean point of view. It does not set, it does not sink into the waters, and it is never hidden, but goes around the world, running through the angles of the sky, giving life to the land and nourishing all the hollows with the food of its light, yet to certain regions in turn the bright face of Phoebus is denied... .

(662) When [the sun] follows its southern orbit and the light slopes toward the Atlantic axis, so that its fire is brought to the

farthermost Hyperboreans, bringing itself back to the Achae-
menian rising, it bends with its curved course toward a separate
part of the heavens and passes the turning post. When it is
denied to us, black night rushes from the sky and blind darkness
immediately covers us, but a bright day illuminates those who
are stiffened by exposure to the north. But then, when the shade
of light is in the Arctic, all our people experience a splendid day.

(674) The river then flows from its source through the Tylangi,
through the Daliterni, and through the crops of the Clahilci and
the fields of the Lemenici. These are harsh words, all of which
wound the ear, but they are not to be omitted, because of both
your eagerness and our attention. It then bends ten times with
the wandering spread of its waters. Many report that a sluggish
swamp inserts itself, which the old Greek custom called Accion.
It passes with swift waters through the surface of the marsh.
Again constricting itself into the form of a river, and from there
looking toward the Atlantic waters, our sea, and the west, it
flows forth and cuts the spreading sand with five mouths. Here
the city of Arelatus rises up, in a former era called Theline by its
Graecian inhabitants. Many reasons have propelled my pen to
describe extensively the Rhodanus, but my mind will never be
influenced to assert that the river is the boundary between
Europe and Libya. Phileus, although ancient, thought that the
inhabitants said this. Let this barbaric ignorance be disregarded
and derided, and censured with an appropriate name. The journey for a boat is two days and two nights.

(700) Then there are the Nearchian people and the town of
Bergine, the fierce Salyes, the ancient town on the Mastrabalan
lagoon, and a promontory with a high ridge that the locals call
Cecylistrium. Then there is Massilia itself. The site of the city is
as follows: a narrow entrance extends through the waves, with
the waters touching its sides. There is a lagoon around the city
and the waves lap the town. The city spreads out its inhabited
area so that it is almost an island, and thus the entire sea pours
its power onto the earth. The diligent work of its early founders
skillfully and completely defeated the shape of the place and its
natural fields. But it would be pleasing to reduce these ancient
names to the new ones...[713].

Commentary

Lines 1–31

The first 79 lines of the poem are a prologue.[25] This is in the form of a dedication to Probus, either Sextus Petronius Probus or his son Amicius Petronius
Probus, who were active in the middle to late fourth century AD. The Tauric

Sea is the modern Sea of Azov, the northern extension of the Black Sea and the farthest extremity of the Mediterranean system. The ancient region of Tauros, roughly the modern Crimea, had long been famous in Greek mythology because of its association with Iphigeneia. Avienus may have cited the Tauric Sea merely to demonstrate that Probus' geographical curiosity went far: Avienus pointed out that learning about this remote region would be a basic part of his education, something that he would remember in his old age.

In order to educate Probus, Avienus had consulted old pages (*vestustis paginis*, line 9). There is a sense of mystery in his research, since through his access to this material he provided information not generally available (*secretiore*, line 11). But Probus was well positioned to absorb this obscure and ancient material (the "secret of things," line 22). He had approached Avienus and asked for this sort of information, a recognition of Avienus' published expertise in geographical matters, since he had already produced his *Descriptio Orbis Terrae*. Probus' interest in the Tauric Sea demonstrates that the *Ora Maritima* did cover that region.

Lines 32–41

One may doubt that a question about the Maeotic (Maiotic) Sea (the same as the Tauric Sea of line 2) generated an entire treatise, but it demonstrates Probus' broad interest in geography. Authors since Herodotus had discussed the sea[26] but Avienus chose as his example a Roman author who was perhaps the earliest to mention it in Latin. C. Sallustius Crispus had a problematic political and military career in the 50s and 40s BC, and then withdrew from public affairs and wrote historical monographs. There are no certain surviving fragments of any writings on the Maeotic Sea, but he did include a detailed account of the Black Sea and its environs as part of his consideration of the wars with Mithridates VI of Pontos in the early first century BC: enough quotations survive to assume that his report on the Maeotic Sea may have been a topographical digression.[27] Mention of Sallust also gave Avienus a staging point for his forthcoming list of Greek sources, a subtle way of giving a Roman authority precedent and priority, reinforced by his eulogistic comments about Sallust's style and abilities as a historian.

Lines 42–50

Eleven sources are listed, perhaps taken from a Hellenistic geographical work of unknown authorship. Thucydides, active into the early fourth century BC, is the latest who can be dated with certainty, but some of the obscurities on the list may be more recent. Hecataeus (Hekataios) of Miletus wrote a *Periodos Ges* (*Circuit of the Earth*) around 500 BC, of which many fragments remain.[28] In doing so he established the genre of Greek ethnographical writing and created one of the prototypes of geographical scholarship. Hellanicus

(Hellanikos) of Lesbos, active in the fifth century BC, wrote extensively on mythology and ethnography, and numerous fragments survive.[29] Phileus (Phileas) of Athens was a little-known geographical writer of the fifth century BC.[30] Scylax (Skylax) of Caryanda was commissioned by Dareios I of Persia at the end of the sixth century BC to sail down the Indos River, and eventually reached the head of the Red Sea. He wrote a report of his cruise as well as other geographical accounts.[31] Pausimachus of Samos is otherwise unknown, but if the names in the catalogue are approximately chronological, he lived in the fifth century BC.[32]

Damastus (more properly Damastes) from Sige (actually Sigeion, in the Troad) was said to have been a student of Hellanicus, and thus was active in the mid- to late-fifth century BC. He wrote extensively on history and geography.[33] Bacoris (Bakoris) of Rhodes is otherwise unknown, and may have been Egyptian in origin, perhaps from the early fourth century BC.[34] Euctemon (Euktemon) of Athens was an astronomer of the fifth century BC, but Avienus is the only extant source to consider him also a geographer, and there may be a confusion with an otherwise unknown Euctemon of Amphipolis (line 337).[35] Cleon (Kleon), from Syracuse in Sicily, wrote a work titled *On Harbors* that was occasionally cited by later authors. He may have been active in the late fourth or early third century BC.[36] Herodotus and Thucydides need no further comment, although it is interesting that Herodotus was said to have been from Thurii (Thourioi), the pan-Hellenic city in southern Italy that was established in 443 BC and where the historian spent his last years.[37] Except for Herodotus and Thucydides, listed at the end for emphasis, the catalogue is probably roughly chronological, bracketed by Sallust at the beginning (to emphasize the Roman outlook) and ending with the two greatest Greek historians of the classical era. Yet few of these 11 names appear elsewhere in the extant treatise, indicating that the list was merely copied by Avienus from an earlier source, probably of the third century BC or later, without reference to how the individuals were actually cited.

Lines 51–79

Having completed his list of sources, Avienus then provided a catalogue of the topics included in the *Ora Maritima*. The account is highly poetic, and is probably his own creation rather than an extensive adaptation from an earlier source, although there are numerous vestiges of an underlying *periplous*.

The Tartessian Strait is a rare and perhaps early term for the modern Strait of Gibraltar. For Tartessus (Tartessos), the southwestern part of Iberia, see line 85. The poetic image describes the Atlantic Ocean thrusting through the straits into "Our Sea" —the Mediterranean—with its effect felt throughout the Mediterranean system. A lengthy list of phenomena follows, including safe harbors, rivers entering the sea, and islands rising up, all information for

139

sailors. The entire account is from an oceanic perspective, in the manner that the shore would be visible from a ship.

Avienus revealed that the geographical limit of the *Ora Maritima* would be Scythia (the north shore of the Black Sea) and the sea (the Euxine) itself. The name Euxinus was the standard term for the Black Sea. There is also a reference to Avienus' own adaptation and translation of the *Periegesis* of Dionysius Periegetes, indicating that it had been completed before the *Ora Maritima* and that the present work was to some extent an abridgement of the earlier one.

As a final point, Avienus concluded his introduction by returning to Probus, expressing his hope that his protégé would find the work useful (lines 74–9), but again referencing the somewhat mysterious quality of the information presented.

Lines 80–93

The size of the earth had been determined in the second half of the third century BC by Eratosthenes of Kyrene in his *Measurement of the Earth*, who realized that it was much larger than the known portions.[38] Since earliest times it had been understood that the landmass of the earth was surrounded by the External Ocean: Homer had some sense of the concept, although imperfectly defined, and by the fourth century BC the idea of an encircling Ocean was taken for granted.[39]

Avienus' perspective is from the western end of the entrance to the Mediterranean, where a sailor would see the Atlantic Gulf opening out before him. The term "gulf" (*sinus*) was probably used to avoid confusion with the more generic term Ocean, allowing Avienus effectively to limit himself to the modern Gulf of Cádiz, bounded by the southern coast of Iberia on the right (north) and the northwest African coast on the left, totally a seaman's perspective.

Gadir is modern Cádiz (lines 267–9). Tartessus is more a regional term; the evidence that it was also the name of a city is scant and disputed, but it would have been easy for the toponym to have both meanings. The earliest Greek reference reports that it was either a kingdom or a trading district.[40] The Columns of Hercules (Pillars of Herakles) are defined in many ways, but the two peaks of Abila and Calpe (Kalpe) are the most common location (lines 343–9). The fact that the strait would be subject to the north wind would be of use to sailors passing through, but this may merely be a misplaced reference from a more northern location, perhaps related to the Column of Baria (emended to Boreios) of the *King Nikomedes Periodos*.[41]

In an abrupt shift (line 90), the text discusses the ridge of Oestrymnis. This is a location at the northwestern edge of Brittany: the toponym varies in the sources but in all cases originated with Pytheas, the Massalian explorer of the second half of the fourth century BC. The name survives in modern Ushant (or Ouessant), an island off the northwestern end of Brittany that was probably connected to the mainland in antiquity.[42] It has long been a major sailing point for those coming from the English Channel into the Atlantic, where they would first be exposed to the *notos* (south) wind.

Mention of the Oestrymnis Ridge marks the beginning of Avienus' actual *periplous*, an itinerary leading back to Massalia. The transition is obvious, although perhaps initially baffling to the reader, representing the way that Avienus chose to handle the material. There is no reason to believe that there are any textual or topographical errors at this point. His use of the phrase "a more ancient era" demonstrates the change from his own world, and that of Probus, to that of the underlying *periplous*. Mention of the "warm *notos*" also indicates a cultural transition: a sailor would be heading from the hostile cold of the north to a warmer and more familiar world.

Lines 94–107

Before the narrative of the *periplous* moves south, there is an ethnology of the Oestrymnis region and a discussion of some places beyond it, presumably toward the north. The Oestrymnic Bay is probably the modern Bay of Douarnenez, between Brest and Quimper, which would have been more extensive in antiquity when Ushant was connected to the mainland. The Oestrymnides Islands are presumably some of the many in this region, but again changes to the coastline make specific determination impossible.

Tin (*stannus*) was a well-known product of this region, and may have been the reason for the original Greek interest.[43] Somewhere in this area were the mysterious Kassiterides (a toponym based on the Greek word for tin), one or more islands that were known to the Greek world from at least the fifth century BC, and probably had been discovered previously by a certain Midakritos, presumably a Phokaian or Massalian.[44] Tin was worked in Cornwall from the late sixth century BC. Interest in lead (*plumbum*) seems more a Roman matter, and mention of it may be Avienus' addition, or a reference to "white lead" (*plumbum album*), another name for tin.[45]

One of the more detailed ethnologies of the *Ora Maritima* follows, probably taken directly from an early source. Part of it is typical idealization of peoples on the edge of the known world, but there are certain autoptic details provided by Greeks who encountered the locals. There was particular admiration for their seamanship, for unlike Greeks they had to sail regularly on the External Ocean, and, moreover, in boats that seemed flimsy by Mediterranean standards. The Ocean was seen to be infested with sea monsters (*beluae*, a generic Latin word for any kind of large threatening animal). These were probably whales, occasionally encountered by Greeks but little understood. The most extensive report of them was by Nearchos and Onesikritos, who sailed from the mouth of the Indos to the Persian Gulf in 325–324 BC under orders from Alexander the Great.[46]

141

The local boats made from skins were, to Greeks, another oddity of the region. They were light and manoeuverable, and thus versatile in the harsh waters. Julius Caesar was impressed enough to have them built for his own use during the civil war of 49–48 BC. Such boats had a long history in the northwest of Europe.[47]

Lines 108–29

Two days' sail from the Oestrymnides was the Sacred Island (Sacra Insula), a toponym that was anachronistic to Avienus. It is not possible to make a reasonable supposition regarding the distance: there are reports from the Hellenistic and Roman periods of 160 to 270 nautical miles (300–500 km.) for a two day sail, but applying this in any reasonable way to Avienus' data is speculative.[48] Yet there seems little doubt that the Sacred Island is Ireland, mentioned by its ethnic name, the land of the Hiernii, in line 111, a Greek and Latin version of the indigenous name Iweriu. Sacra Insula is a translation of Greek Hiera (or Iera) Nesos, which was probably an erroneous rendering of the indigenous name; Ireland is normally Ierne in Greek.[49] The nearest point of Ireland (in the vicinity of Cobb) lies about 400 km. from Ushant. Noting the richness of the local turf (*caespes*) has an autoptic quality and became part of the Greco-Roman perception of the island.[50]

There was another island near to the Sacred Island, that of the Albiones. This has consistently been identified with Great Britain, and, depending on Avienus' source, this may be the earliest use of the name, perhaps from Pytheas in the late fourth century BC. The Welsh and Scottish coasts are visible from Ireland, and any route there from the Oestrymnides would pass close to the west end of Cornwall (which would look like a relatively small island), but it is not possible to localize the toponym Albion.[51]

At line 113 (through line 129) the thrust of the *Ora Maritima* changes suddenly from early Greek exploration in the northwest of Europe to that of the Tartessians and Carthaginians. This is probably from another source, inserted by Avienus at this point. The Tartessians, the indigenous population of southwest Iberia, had extensive trade and commercial contacts. As biblical Tarshish, they had connections with King Solomon of Israel at the east end of the Mediterranean,[52] and there is no reason to doubt that in the other direction they explored the Atlantic coast of Europe as far as the British Isles, probably seeking mineral resources; their export of tin would mean they had access to northwest Europe.[53] Although there were some tin sources in northwest Iberia, the evidence is uncertain, especially for this early period.[54] Rumors of this trade, whether in Iberia or beyond, may have been what encouraged Greeks to set out in the same direction, probably as early as the seventh century BC.[55]

Of particular interest is the first of Avienus' three references to the voyage of Himilco (Himilko or Himilkon), a Carthaginian contemporary of Hanno.

He is known only through these three citations and a sparse notice in the *Natural History* of Pliny.[56] The limited information on the explorer is a frustrating contrast with the data about his probable relative Hanno. How Avienus gained access to Himilco's report remains a mystery: it is clear that he had some kind of written document (lines 117–19), which he identified as actual Carthaginian records (lines 414–15). Pliny presumably had the same material, which he obtained through a secondary source. It is easy to see Himilco and Hanno in tandem, one going south and the other north, and it is speculatively possible that Greek summaries of both cruises were produced, but that of Himilco vanished at a relatively early date, only surviving in scattered references drawn from an unknown source by Pliny and Avienus. Since the earliest extant documentation of Himilco's voyage was over 500 years after the fact, the chance for corruption of detail is high, but some essential facts can be understood.

His commission was explicitly documented by Pliny: he was sent to learn what was beyond Europe, or beyond the Oestrymnic region (a definition of Europe from his own day, not that of Pliny). It is possible that Avienus' information on the Sacred Island and Albion was derived from Himilco's report, but its tone is different, because there are no toponyms or ethnyms (other than Europe) in any of the material directly attributed to Himilco.

All three citations of Himilco by Avienus are similar: there is no discussion of places or peoples encountered (a distinct contrast to Hanno's surviving report), but an emphasis on the hazards. Some of these are common material about sea voyages to remote places: adverse winds, the eternal dampness, shallows (perhaps due to tidal actions), and sea monsters. Of greater interest is the notice of four months for the cruise, which may be a Carthaginian way of telling competitors that there was no point in attempting to find the places that Himilco visited because they were so remote, thus keeping trading details secret. But it is more likely that it represents the total length of the journey, which, like that of Hanno, would have included stopovers. It is possible that Himilco reached the Azores, which lie about 900 km. west of the Iberian peninsula, and which would account for the four months, but evidence of any ancient knowledge of the islands is questionable.[57]

The other point of interest is the sluggish sea. It has been suggested to be the Sargasso Sea, but this is far west and unlikely, and it is more probable that Himilco meant a known region four days beyond the Columns of Hercules where low tides caused a thickening of the seaweed. Here fishermen from Gadir found excellent tuna that they exported to Carthage, and Himilco may have attempted to find it for himself.[58] The mention of sea creatures swarming around the ships, if taken literally, means that the expedition was large enough to have more than one ship, and taking note of such encounters is certainly a reasonable comment on the peculiarities of oceanic travel from those not generally familiar with it.

Lines 129–45

The tone of the account again changes sharply. Although it is possible that Himilco continues to be the source, the citation of toponyms and ethnyms makes this unlikely. This is a report of one or more expeditions north of the Oestrymnides into little-known and treacherous regions. Lycaon was a mythological personality originally associated with Arkadia in the Peloponnesos, but who came to be connected with the far north when his daughter Kallisto was turned into the constellation of the Bear.[59]

The Ligurians lived in the regions between Iberia and Italy, but there was a memory of their emigration from a northern region, a point of view reinforced at lines 195–8. There is no solid evidence where the original Ligurian homeland was or when the Celts forced them south.[60] Strabo's description of the Ligurians—a primitive people with a subsistence lifestyle—although in the context of their location on the Mediterranean coast, has vestiges of a more basic existence. Their fear of the sea, reported by Avienus, shows that they were once an inland population. Yet his account fixes them where they were in historic times, the mountainous region (essentially the Maritime Alps) above the modern French coast, with eventual movement more toward the shore. The account is from a Massalian perspective and shows that the Ligurians were probably in position before the founding of that city around 600 BC. If Avienus' information is from about this time, this is the earliest reference to the Ligurians, since they did not generally come into Greco-Roman interest until the Romans moved into the region in the third century BC.[61]

Lines 146–57

Returning to where he digressed after line 107, Avienus moved down the west coast of Europe into the Great Bay (Magnus Sinus), the modern Bay of Biscay, at a place he called Ophiussa ("Snaky"). The name, with its ending in -ussa (-oussa), is distinctly Phokaian, and is one of about a dozen such toponyms scattered from Italy west into the Atlantic. Ophiussa is the northernmost reported Phokaian toponym, and is probably misplaced at this point in the *Ora Maritima*, since it should be on the Iberian coast, probably at Cape Roca near Lisbon, the westernmost point of Europe; it is properly located at line 172. But the name also had a broader generic focus, seemingly meaning the entire Iberian peninsula and even points north, perhaps a reflection of the uncertainty of the actual situation of some of these places.[62]

There was an established trade route from the Great Bay to the region of the Mediterranean (Inner Sea) called the Sardus, a common name for its western portion, based on the island of Sardo (Sardinia).[63] There were two possible paths. One was from the mouth of the Liger (modern Loire) River near Nantes, and upstream almost to its source. From there it was only a short portage to the Rhone system and down to the Mediterranean and Massalia. This had been in use since prehistoric times, and had been taken by Pytheas in the fourth century BC. It was probably known to Polybios, and access to it was one of the reasons for Massalian commercial success.[64] Another route would have been from the region of modern Bordeaux up the Garonne and across to the Aude, reaching the Mediterranean around Narbonne. This is shorter than the other one, and may conform better to the reported timing of seven days.[65]

Equating the size of remote geographical forms to the Peloponnesos was an odd feature of Greek geographical thought: the author of the *King Nikomedes Periodos* wrote that the Dalmatian peninsula of Hyllike was similar in size to it. Perhaps the most unusual of such comparisons was Anaxagoras' statement that the sun was in fact greater than the Peloponnesos.[66] Because it is not clear what Avienus meant by Ophiussa at this point, it is not possible to critique the comparison.

The toponym Oestrymnis, although localized in northwestern modern France, also came to have a broader sense of much of far western Europe, perhaps everything north of Ophiussa. Avienus saw the two names as consecutive but topographically identical, and provided an aetiology for the change, but it is more probable that Ophiussa is the Phokaian name and Oestrymnis the Massalian for overlapping territories. Ophiussa would be the earlier, and significantly is Greek, not indigenous; by the time the Massalians arrived, after 600 BC, local names were incorporated into the literary map of the region.

Lines 156–7 are peculiar, expressing the idea that ophids not only drove the locals away but gave their name to the territory (presumably Ophiussa). Although this may merely be an odd aetiology, it is also possible that *serpens* is a misunderstood ethnym, and that the incident is a memory of an early ethnic invasion, long before the era of Greek exploration and settlement.[67]

Lines 158–82

The Ridge of Venus is somewhere on the northern Iberian peninsula, perhaps Cabo Higer at the western end of the Pyrenees, or a point farther west. The two islands are probably the modern Illottes Aguillons, just offshore of Punto de Aguillons, known in antiquity as the Trileukoi and which probably never had any significant habitation.[68] Here sailors could leave the coast and strike northeast, going directly to the northwest corner of France, about 500 km. away, thus avoiding the lengthy route along the coast of the inner Bay of Biscay. This was a rare example of sailing away from the coast in antiquity, but it provided a shortening of several days.

The Aryium Promontory, located five days from the Columns of Hercules, is probably modern Cape Finisterre, where the Iberian coast makes its final turn to the south, another important point for sailors. The toponym may be a variant of the ethnym of the Artabrians (or Arrobrians), who inhabited this region.[69] The timing of five days is plausible; the distance is about 1,000 km.

The island sacred to Saturn, specifically noted to be out to sea (*pelagia*) is one of the Berlengas, about 10–17 km. from the coast from a point north of Lisbon. Saturn was presumably Carthaginian Baal Ammon, indicating that the islands were one of their outposts. They have long been noted for their luxuriant vegetation. Isolated islands are always of interest to sailors, as well as dangerous, and thus needed to be located, yet there are many stories from ancient to modern times about their perils, in a sense going back to Kirke and Kalypso. These are sailor's tales but reflect the problematic nature of landing on remote shores, and the Carthaginians may have promulgated such accounts within their territories in order to keep others away.

At line 172 the Ophiussa Promontory is a specific toponym, not a vague geographical term. It is almost certainly Cabo de Roca, which rises up to 140 m. just west of Lisbon, and which is the westernmost point of Europe. It was an important location for sailors, where ships would need to turn more easterly in order to keep close to the Iberian coast. It is also the northernmost Phokaian toponym: from here south the data available to Avienus became more detailed. The promontory was two days' sail south of the ridge known as Arvium, presumably the Aryium of line 160.

South of Ophiussa was a bay where navigation was difficult: this is probably the small bay at Lisbon. The northern half required the westerly zephyr since a ship, tacking slightly, would head southeast, but this direction would not clear the south end of the bay. Yet it is difficult to explain how the *notus* (south) wind could be of any use when moving southerly, and it may be a misplaced reference from farther down the coast where it turns more easterly.[70]

The land route from the coast left from an unspecified point, probably in the vicinity of Lisbon, and headed southeast toward the region of Tartessos, four days away. The Tartessian Shore is probably that around Gadir: Avienus had already equated the two (line 85). There is no obvious route but it probably kept close to the coast and avoided the rough interior. It continued east to reach the Mediterranean at Malacha (modern Malaga, line 426). The total distance is about 650 km., so it would be essentially impossible in nine days, and Avienus certainly had some misgivings about the timings. But even if inaccurately expressed, the land route was an important piece of information for mariners: it gave them some idea of how far they were from the Mediterranean, and provided a way home should continuing by sea become impossible. Adverse winds could last for days or even weeks, especially in the autumn and winter.[71] Moreover, the land route became important with the rise of Carthage and their control of the Straits of Gibraltar. As early as the late sixth century BC they actively excluded foreign ships from areas that they considered their own, and there was repeated tension between the Massalians and the Carthaginians about the sea route.[72] Development of land routes that kept the Massalians as far away as possible from the Carthaginians would have been a priority.

Lines 182–94

Continuing to view matters from a seaman's perspective, the next feature to rise up was the Cempsician Ridge, named after the local Cempsi (line 200). It was the long promontory (modern Espichel) that marks the north edge of the Bahía de Setúbal, and which rises to 140 m., the next high point on the coast south of the Ophiussa Promontory. The island of Achale has disappeared, or has joined to the mainland, and thus it is difficult to explain its unusual properties, but they may be related to tidal activity or river outflows.

Lines 195–204

The Cempsi seem to have been in the interior uplands of western Iberia, but the only other source to mention them, Dionysios Periegetes, placed them closer to the Pyrenees.[73] This may reflect a movement of the population from early Greek times into the Roman period. The Sefes would have been their neighbors but are otherwise unknown. The ethnym Ligians (Ligus in the text) is an emendation from the incomprehensible *lucis*; if Ligus is correct the word may be related to the Lusitanians. The Draganes are also otherwise unknown but their location, where they were affected by the "snowy north wind," suggests the rugged interior of Iberia. The scattered nature of these ethnyms—a feature of the *Ora Maritima*—may be due to ethnic information of uncertain origin reaching the Greek cities on the coast.

The island of Poetanion may be modern Alpeidão, in the Tagus estuary, which seems to preserve the ancient name (with the addition of the Arabic *al-*), where the Sefumi lived and mariners would find a decent harbor.[74] For the Cempsians, see line 195. The Cynetes (Kynesioi or Kynetes) had long been noted as the most western population in Europe. They would have lived in the far southwest of the Iberian peninsula. The Cynetician Ridge would be around Cabo de São Vicente, the corner of the peninsula, generally called the Sacred Promontory.[75] It was not the westernmost point of Europe—Cape Roca west of Lisbon is over half a degree of longitude farther west—but it was a conspicuous topographical and navigational location and would seem to be a westernmost point, extending, as Avienus said, toward the setting of the stars. Ships from the north would round it and then have a direct easterly sail to Gadir and the Columns of Hercules. Avienus' poetic excursis on the feature emphasizes its importance.

Lines 205–23

The Cynetes occupied a broad region of southwest Iberia, and thus Avienus placed their eastern boundary beyond or at the Ana (or Anas) River, the modern Guadiana and the boundary between Portugal and Spain. It is one of the major streams of the Iberian peninsula, flowing 815 km. across south-central Iberia and turning south before emptying into the Atlantic through several mouths, reflecting Avienus' "two rivers." It is sluggish in its lower course, creating a fertile floodplain.

The account returns to the vicinity of the Cynetician Ridge (line 201), with reference to a local shrine of Saturn (Phoenician and Carthaginian Baal Ammon), located perhaps at modern Ponta de Sagres. The two islands—one unnamed and the other called Agonis—were along this coast but cannot be located and have probably been joined to the mainland. The use of goat hair for military tents and naval clothing was a regular feature: Avienus' line 220, obviously his interpolation, is a close paraphrase of Vergil, *Georgics* 3.313. The sail of one day from the Cynetician Ridge to the mouth of the Ana (about 150 km.) is reasonable.

Lines 223–40

The *Ora Maritima* has returned to the region of Tartessos, used in a general sense for the south-central part of Iberia (lines 85, 113, 179). Tartessos was an ancient name for the Baitis River (modern Guadalquivir), which flows for 657 km. across southern Iberia and in antiquity emptied into a large estuary, now mostly reclaimed, west of Gadir. The name Tartessos for the river was documented as early as the sixth century BC, and reflected early Greek (or Phoenician) use of the regional toponym for the river.[76]

The Zephyrus ("West Wind") Ridge and the citadel of Zephyris suggest localities exposed to the west, with onshore winds creating foggy and humid conditions. This may be the region of the Serra do Caldeirão in southern Portugal, a range that reaches the sea near modern Tavira and Faro, and which was a region of early Phoenician settlement.[77] Favonius is the Latin word for the Greek zephyr, which would effectively drive a ship toward the Columns of Hercules and the Mediterranean.

Lines 241–59

Atlantic fogs (alluded to at lines 228–30) were an anomaly to Mediterranean peoples, and helped to create an impression that the coastal regions were associated with the dark and gloomy underworld. The similarity of the name Tartessos to Tartaros, in the depths of the underworld, coupled with the fact that the far west was connected with the sunset and thus darkness, led to the assumption that Tartessos could indeed be the location of the underworld.[78] Expectedly, data were accumulated to prove this point of view. The ridge of the infernal goddess is probably the shrine to Phosphoros (a cultic name of Hekate) mentioned by Strabo.[79] Other mysterious cultic centers were in the region, including a hidden cave and an invisible sanctuary. In fact, this entire portion of the *Ora Maritima* (lines 241–7) has a Homeric tone, although it is impossible to determine whether this is Avienus' creation or something taken from earlier sources who put their own Homeric outlook on the topography.[80]

The names Erebea, or Etrebea (a regional toponym), Herbi (a town name), the ethnym Herbi, and the Hiberus River are all linguistically connected and refer to the district around the estuary of the modern Río Tinto, about 40 km. east of the mouth of the Ana. Although certainly indigenous names, they reminded Greeks of Erebos, a term for the underworld mentioned regularly by Homer, giving further credence to the idea that this region was in some way connected with the underworld.[81]

The town of Herbi had long been abandoned when Avienus or his source wrote, and may have been an early seaport for the Tartessians. The Hiberus River is the modern Río Tinto, which flows from the rich mining district northwest of Seville and reaches the ocean in 100 km. It was probably the means of exporting the Tartessian mineral wealth. Because there was another and more famous Hiberus River (the modern Ebro) in the north of Iberia (which actually gave its name to the peninsula, although Avienus was unaware of this), and which was the second longest in the peninsula (930 km.), Avienus was careful to point out that the Hiberus at Herbi was a different stream. The Vascones lived north of the upper part of the northern Hiberus, and were unknown to the Greek world and not encountered by the Romans until the campaigns against Q. Sertorius in 76 BC. Thus mention of them is an intrusion into the narrative of the *Ora Maritima*, either added by Avienus for clarity, or from a source that was oriented on the first century BC.[82]

Returning to the south of Iberia, the *Ora Maritima* cited the local Hiberi who lived west of the Hiberus River, with the Tartessians to the east. For the Cilbiceni, see line 303. Cartare is probably one of the islands in the Tartessos River estuary, below modern Huelva. Today this region has changed due to siltation, but the two mouths were still apparent in the second century AD.[83] For the Cempsi, see line 200.

Mt. Cassius is probably the Asperillo Cliffs on the coast southeast of Huelva, a national park today with striking formations that would have been conspicuous to sailors. But the toponym is peculiar, since it appears to be in the form of a Roman *nomen*, not impossible but unusual in southern Iberia. It may be a latinization of Kasion, the name of two conspicuous mountains in the eastern Mediterranean. One is on the Egyptian coast east of the Delta (modern Ras Qarun), an important point on the route between the Levant and Egypt. The other (modern Jebel el-Akra) is in Syria south of Antioch.[84] It is remotely possible that the name of one of these mountains was transferred to Iberia. Yet Avienus connected the toponym with *cassiterum* (*kassiteros*), the Greek word for tin, a product that the Tartessians may have traded. None of the explanations seems particularly credible, and Cassius may merely be a latinized indigenous toponym.

149

Lines 260–83

The citadel of Geron was somewhere near the outlet of the Tartessos River, probably on the eastern side. This is an area that has changed considerably since antiquity, and no further locating is possible. Avienus implied it was a local toponym that had become part of Greek myth, since Geryon, the three-headed and three-bodied giant, was early associated with this region, the object of the tenth labor of Hercules. The story was ancient, mentioned by Hesiod,[85] and the subject of an epic by Stesichoros, the *Geryoneis*, of which a number of fragments survive.

The end point of the journey of one day is vague, but it certainly would be an easy timing from the Tartessos estuary to Gadir. Gadir (from Phoenician 'Gdr; Gadeira in Greek, and generally Gades in Latin; modern Cádiz) was said to have been the original Phoenician settlement in Iberia, allegedly founded around the time of the Trojan War but more probably in the eighth century BC. The etymology "fenced place" is accurate, indicating an original fortified trading post, which may have been called Tartessos after the surrounding region (line 85).[86]

The comment about its poor and small status is certainly Avienus' addition; it was wealthy and prosperous into the first century AD and probably beyond.[87] Avienus, in a rare personal comment about his travels, visited the town in the mid-fourth century AD and found it—perhaps writing in exaggeration— "a heap of ruins." The Temple of Hercules (Melqart) was probably the most prominent feature in the decayed town.

In fact, the entire section of lines 273–84 is Avienus' insertion and reflects his personal experiences in the city. Juba (II) was the scholarly king of Mauretania, who reigned from 25 BC to AD 23 or 24. Married to Cleopatra Selene, the daughter of Cleopatra VII and Marcus Antonius, who helped implement his research, he was a major scholar particularly noted for his ethnologies on the southern part of the inhabited world, from the Atlantic to India. The royal couple renovated the decayed Carthaginian trading post of Iol (modern Cherchel in Algeria) and turned it into an architecturally innovative and culturally prominent royal capital named Kaisareia. The king was honored throughout the Roman empire and held the office of *duovir* in both Gadir and Carthago Nova (modern Cartagena): this was the chief magistrate in a Roman *colonia* or *municipium*.[88] The date is uncertain beyond the parameters of the half century of Juba's kingship. His office is only recorded by Avienus, who may have seen the information on an inscription. There is no specific explanation as to why the king was chosen by the locals for the office, although presumably it was due to recognition of his cultural and political prominence in the western Mediterranean.

Lines 283–98

Avienus returned to material obtained from his earlier literary sources. The Ligustine Lake is probably part of the Tartessos estuary, since there is no lake farther upstream. The name may have been corrupted and confused with the Greek word for the indigenous population of the region of Liguria in modern France and Italy. A later reference to a city named Ligystine near Tartessos may only have been derived from Avienus.[89] The river opens out into a broad estuary that even today has many channels.

Although Mt. Argentarius (Silver Mountain) may have been a specific location, the name is more likely a regional term applied to the rugged interior north of the Tartessos River, the modern Río Tinto district, which was a significant mining area until modern times. The name, although seemingly Greek, is probably indigenous, from Celtic *argant*, or silver.[90] In 630 BC Kolaios of Samos became the first Greek known to have reached Tartessos and bring its wealth to the eastern Mediterranean. Eventually relations were established with the local ruler, King Arganthonios (The Silver King), who made contact with Phokaia and helped finance their defense against the Persians.[91]

Stagnum is a mixture of silver and lead, indicative of the nature of the raw ore. The poetic description of the mountain has the flavor of an idealized account of the wonders of a remote land, a feature of exploration and travel literature into modern times. Even today the Río Tinto region has an unworldly aspect.[92] There certainly would have been some alluvial minerals washed down from the mountains, but most of the mining in this region was through excavation, and numerous pits and shafts from antiquity are visible today.

Lines 298–317

The following section describes populations in the vicinity of Tartessos, and perhaps inland from there. The Etmanians, Cibiceni, and Ileates are otherwise unknown. For the Cempsi, see line 195. The second reference to the citadel of Geron (see line 263) may help locate some of these peoples, but the narrative is rather scattered. Here Mt. Tartessos seems to refer to the ridges southeast of Gadir (see lines 317–18). The river is perhaps the Besilus or Cilbus of line 320. Avienus may have ineffectively blended sources in this section. He evidently was unaware that Erythia (Erytheia), a semi-mythical toponym, was most frequently placed at Gadir.[93] To be sure, Erythia was located in a number of places, including on the upper Baitis River,[94] but the description provided by Avienus—that it was a Carthaginian settlement on an island five stadia (1 km.) from the mainland—fits the location of Gadir. The island of Venus Marina with its temple to the goddess (Ashtart or Tanit) is the small islet of San Sebastián, now part of the mainland.[95]

Lines 317–40

The account continues to the southeast along the coast from Gadir to the Columns of Hercules. The unnamed mountain is Mt. Tartessos (line 308), here more firmly placed in the uplands east and south of Gadir. The Besilus

River is probably the modern Barbote, a short stream that empties into the Atlantic about 50 km. southeast of Cadíz. The Cilbus, perhaps in the region of the Cibiceni (line 303) cannot specifically be located, but would be one of the several other streams on this coast, perhaps the Salado.[96]

The Sacred Ridge is presumably modern Cape Trafalgar, the only promontory on this coast. Herma cannot be identified, despite the specifics, and is not mentioned in Strabo's detailed account of this coast. The name is suspiciously close to "Hera," who was associated with this region, but no lake is known (see also line 336).[97]

At line 326 there begins a discussion of the Columns of Hercules (through line 418), unusual for the *Ora Maritima* because of its length and the citation of several sources by name. Presumably this was taken from a Hellenistic regional account, which contained early *periplous* antecedents but was more focused on the region around the Columns. Moreover, the narrative is not limited to the Iberian side, but includes material from Libya (Africa) as well.[98]

To some extent these lines are an enclosed separate narrative, which has probably been thoroughly rewritten by Avienus and somewhat romanized (lines 346–7). It begins with mythology and how Hercules (Herakles) created the straits between Europe and Libya in order to drive the cattle of Geryon through them. Here Herma is the straits themselves, 108 miles (160 km.) long and three miles (4.4 km.) wide. The length is actually that from Gadir to the Mediterranean, something navigators would want to know; the width is far too narrow and the number may be corrupt. Two sources were cited by Avienus, but they may be limited to the specific comments, not the general discussion. Dionysius is probably Dionysios Periegetes, but the remark attributed to him does not appear in his extant treatise, and Avienus may have been referring to his own edition of the work. Euctemon is probably the scholar from Athens, of the fifth century BC (line 47), although there may have been another personality of the same name from the Macedonian town of Amphipolis. For the Sacred Promontory, see line 322. It was seen as the west entrance to the straits.

Lines 341–9

The Columns of Hercules (Herakles) were at the entrance to the Mediterranean and also divided Europe and Africa (Libya). Homer called them the Pillars of Atlas, but by the early fifth century BC they were attributed to Herakles and seen as the end of the world.[99] Avienus had already mentioned the two mountains, Abila and Calpe, that were thought to mark the columns (lines 87–8), but this portion of the treatise is more romanized than usual. "Hispanus," is a Latin term not documented before the first century BC.[100] "Punic" is also purely Latin, derived from "Phoenician" but referring to the

Carthaginians, in common use from at least the second century BC. The contribution of Plautus—the Roman playwright of the early second century BC—may be limited to calling Latin a barbarian language rather than any topographical details.[101] The meaning of Abila, "high place," is certainly appropriate (it is Jebel Tariq in Morocco), but has no known linguistic support, at least in Latin, although it may be the indigenous meaning. Calpe (Kalpe, modern Gibraltar) is indeed a type of pitcher in Greek, and thus the toponym may have come from that language.[102]

Lines 350–69

Euctemon of Athens is probably, but not certainly, the scholar from Amphipolis of line 337. He provided an idiosyncratic view of the Columns of Hercules, locating them on two islets, probably modern Palomas near Algeciras and Peregil off the African coast, an identification also found in the *King Nikomedes Periodos*, although to be sure, the islets are more than 30 stadia (6 km.) apart.[103] Avienus' account has returned to the *periplous* format with an emphasis on the navigational difficulties in the region of the straits, including shallows (which exist at the west end) and various currents. Luna Island cannot be located, but there were restrictions on approaching the shrines of Hercules on the two unnamed islets, and this may have had something to do with the political dynamics in this region of Carthaginian, Greek, and eventually Roman competition.[104]

Lines 370–89

Two further sources are cited. Damastus (actually Damastes) and Scylax. Damastus' figure, if it is either width or length, is quite inaccurate. The length depends on how it is measured but the width is about 24 km., not the 1.4 reported. The error may be explained by the report attributed to Skylax, comparing the strait to the Thracian Bosporos, which is generally seven stadia across. It might be that there was confusion between the perceived entrance and exit to the Mediterranean.[105]

There were numerous Carthaginian settlements on the Mediterranean coast of Iberia, but as the Romans spread through the region in the late third century BC, these were assimilated and refounded as Roman outposts. Nevertheless a Carthaginian cultural presence would have long remained. The Carthaginian ships described would have been coastal boats suitable for entering the shallow estuaries, not their larger and more study ocean-going vessels.

At line 383, Himilco, the Carthaginian explorer of the region north of the Columns of Hercules, was introduced for the second time (see line 117). The passage is intrusive and has little relevance to the coastal regions around the Columns, other than the explorer needed to pass through them to begin his cruise, and Avienus was careful to make a transition by referring to the "open sea to the west." The account is somewhat loosely expressed since Avienus noted that no one had come to the regions so carefully described. The emphasis is on the problematic winds and fog, repeating the earlier material from Himilco, and its tone may reflect Carthaginian discouragement of others entering this region. It also allowed Avienus to make a general statement about the Ocean, which follows.

Lines 390–416

Avienus' digression on the Ocean is largely taken from Dionysios Periegetes and is a condensation of lines 26–67 of his work, which Avienus had already translated into Latin. The emphasis is on what they called the four great parts of the Ocean, a point of view also advanced by Strabo.[106]

The External Ocean is seen as the nourisher of the Mediterranean ("the inner salt water," an unusual phrase). The description of the Mediterranean is vague, since it is expressed in terms of the penetration of the Atlantic by means of its tides and saltwater. The second gulf is the Hyrcana or Caspian. These were names that were essentially interchangable and reflected different ethnic groups on its shores. Early Greek geographical writers knew that it was an enclosed sea, but the companions of Alexander the Great decided it was an inlet of the External Ocean, a manipulation of topography that served the king's needs.[107]

Avienus' third gulf is defined as extending from India to Persia, in other words, the modern Indian Ocean (not mentioned as such by Dionysios) and Persian Gulf. Finally, there were the "Arabian waters" and the Red Sea, or what was west of the mouth of the Persian Gulf. Belief that the Hyrcana or Caspian was a general term for the Ocean is probably a misreading of Dionysios, where the Caspian is mentioned between the Persian and Arabian Gulfs.[108] Line 403 is a verbatim repeat of 399 and should be deleted as a copying error; there was no belief that "Caspian" was a general term for the Ocean.[109] The description that follows in the *Ora Maritima* (lines 412–15) is the third and final reference to Himilco and the dangers of the open sea, but with the important detail that there once existed a published report of his cruise, expressed more definitely than elsewhere in the treatise, although there is no evidence that Avienus consulted it directly.

Lines 417–24

Avienus told his readers (line 416) that he was returning to his linear *periplous*, abandoned at line 380. The Libystidic Column is the southern Column of Hercules (modern Jebel Tariq), but the form of the toponymic adjective is anomalous, perhaps for metrical reasons. The Chrysus River is the modern Guadiaro, also known in antiquity as the Barbesula, which was the indigenous name.[110] The river enters the Mediterranean at the head of the Bay of Algeciras. The name Chrysus ("Gold") references the Greek impression, perhaps accurate, that alluvial gold was carried down it.

Libyphoenician was a hellenized term that described the Phoenician (later Carthaginian) settlements in northwest Africa, and to some extent in adjacent Iberia (as here). It is first documented in the account of Hanno.[111] Characterizing them as "fierce" shows a Greek bias in the account, presumably from the Phokaians or Massalians. The Massieni (or Mastianoi) were the indigenous people of southeast Iberia, whose center was the town of Mastia (or Massiena), located near the future site of Carthago Nova, at modern Cartagena (lines 451–2).[112] For the Cilbiceni, see lines 255, 303. The Tartessians have frequently been mentioned throughout the *Ora Maritima* as the dominant ethnic group of southern Iberia. The Calactic Bay, whose name may be a latinization of Greek Kale Akte, or Beautiful Promontory, was the western limit of the Tartessians and thus should be the estuary of the Tartessos River (lines 223–4) at modern Huelva.[113]

Lines 425–48

The *Ora Maritima* continues along the southeastern coast of Iberia. The Barbetian Ridge is one of the promontories east of the Columns of Hercules, perhaps in the vicinity of modern Punta de Calaburras, where the coast makes a sharp turn to the north. The Malacha River is the modern Guadalmedina, a short stream entering the Mediterranean at modern Málaga (the homonymous city of line 426). The city was a Phoenician and then Carthaginian outpost which came under Roman control in the late third century BC. It is not the same as Menace (see also line 431), or Mainake, at modern Torre del Mar, which was about 30 km. to the east and was a Phokaian and later Massalian settlement: Malacha and Menace were competing posts for the African trade and marked the boundaries of their specific cultural spheres, although the distinction between the two is far from clear.[114]

The island of Noctiluca and mention of the Tartessians suggests a reversion to points farther west, since there are no known islands in the vicinity of Malacha and Menace, although one may have disappeared. Noctiluca may be one of the islands around the Bay of Algeciras, most likely the island of Luna mentioned at line 367, since Noctiluna is one of the epithets of that divinity, whose primary shrine was at an unknown location on the Palatine in Rome.[115] Such a shrine on the Iberian coast would only have been possible after the arrival of the Romans in the late third century BC. At this point Avienus realized that Menace and Malacha were not the same.

Mt. Silurus (or Solorius) is the modern Sierra Nevada, the isolated and precipitous range rising to 3,479 m. just east of modern Granada. It is only about 30 km. from the sea, and the Greek name for its pine forests (evidently no longer existing by the Roman period) was something like Pityoussa, probably applied by the Phokaians; the toponym appeared in a number of places in the Greek world with such flora (line 470).[116] The west entrance to the modern Golfo de Almeria—where the coast "falls back"—is modern Punta Sabinar, named after a type of local juniper. The Shrine of Venus (Aphrodite) is probably at modern Cabo de Gato, at the east side of the bay, an important navigational point, and the ridge of Venus is probably the uplands stretching to the northeast.

The emphasis is on the deserted nature of this coast, which may be an exaggeration but reflects the retrenchment of Carthaginian power after the middle of the second century BC. Similar circumstances occurred along the Atlantic coast of Africa.[117] The Herma of line 444 is not the same place as the homonym of lines 323–6, as this Herma is in Libya (Africa). From the summit of the ridge of Venus (at 435 m.) one could allegedly see across to Africa, about 150 km. away; the ridge is the southernmost point on this stretch of coast. The location in Africa is probably Cap de Trois Fourches in Morocco, a long promontory extending into the sea that was an important navigational marker.

Lines 449–71

The Namnatian harbor and the city of Massiena (see line 422) are in the vicinity of modern Cartagena, where New Carthage was founded in the late third century BC.[118] Thus this portion of the *Periplous* was based on material gathered before that date. It has remained an important town from early until modern times, due to its particularly fine harbor. The Trete Ridge may be modern Cabo de Palos, where the coast turns sharply north; the name may be Greek, meaning "perforated," because of caves in the cliffs. The island of Strongyle, another Greek toponym ("Round") should be one of the several small islands to the north of the cape.

The extensive marshy area is still visible north of Cartagena, although largely reclaimed as the Campo de Cartagena. The Theodorus River is not a Greek personal name, but a hellenization of the indigenous name, Tader, the modern Segura.[119] It is the most important river of this region, whose mouth is about 65 km. north of Cartagena. Avienus may have expressed incredulity at the frequency of Greek toponyms in this region, but as one traveled along the coast and moved farther from the Carthaginian sphere and into the Phokaian and Massalian one, this was to be expected. He may also have been surprised at the survival of Greek names after hundreds of years of Roman rule. Moreover, the *Periplous* becomes more detailed now that it is within the Greek area. The limit of the Tartessian region was reached at the city of Herna (line 463), which no longer existed at the time of Avienus' source. It may have been at modern Guardamar de Segura, just above the mouth of the Segura River.[120]

After a sandy region, there are three islands which are not named in the *Ora Maritima* and cannot be identified today. They could be the Balearics, well offshore but mentioned at line 471. This region was seen as the limit of the Tartessian territory, a point of view perhaps reflecting the Greek penetration into the area. The description is somewhat confused, since details about the Gymnetian (or Balearic) Islands, which lie about 100 km. offshore, have been mixed into the Iberian coastal account. The Sicanus River was near Sicana (line 479); the Alebus is not identified. But it is unlikely that the Gymnetai (so named because of their habit of going naked in the summer), the inhabitants of the Balearics, had any presence on the mainland.[121] Pityussa ("Piney"), perhaps a Phokaian name, is modern Ibiza; the Gymnesiai proper are modern Mallorca and Menorca. There was no island specifically called Gymnesia.

Lines 472–89

Returning to the mainland, Avienus used the ethnym Hiberi for the population of northeastern Iberia, whose territory extended from the mainland opposite the Balearics to the Pyrenees. Ilerda cannot be located (it is unlikely to be the town far to the northwest at modern Lleidia, usually identified as ancient Ilerda), and was probably near Hemeroscopium, which was a sanctuary and coastal lookout, as well as a Greek emporium. It was where the coast thrusts sharply to the east toward the Balearics, and was a Phokaian shrine to Artemis that became romanized and sacred to Diana. Today it preserves the ancient name as Dénia.[122]

Sicana and the Sicanus River (see line 469) are north of Hemeroscopium; the river may be the modern Júcar (or Xuguer), with the town slightly inland at modern Sueca. The Tyrius River is the modern Turia, the next river north, which empties into the Mediterranean at modern València, where the town of Tyris probably was.

A rare ethnic digression (for this portion of the *Ora Maritima*) discusses the Berybraces.[123] They inhabited these regions and were noted for their cultivation of livestock, although the author of the original ethnography could not resist noting their fierce and wild nature. Presumably they came to the coast in order to trade with the Greeks.

Lines 489–503

The Crabrasia Ridge, near a city of the same name, is probably one of the uplands that come close to the shore north of València, perhaps near modern Sagunto.[124] The Onussa Cherronesus ("Donkey Peninsula") would be to its north, but there is no obvious feature in this region. There was also a town of Chersonesus in the area.[125] The marsh of the Naccarares may be the Albufera region just south of València, but if so it is out of order, and could have been farther north. It held a shrine to Athene (Minerva), but the island is not visible today. Olives remain a major local crop.

Hylactes, Hystra, and Sarna were coastal towns south of the Hiberus River which have not been located. Tyrichae is probably at Dertosa (modern Tortosa), the important lower crossing of the Hiberus River, situated about 15 km. inland.[126] It was the major local market town and a trading emporium for goods coming down the river. The name "Hiberus" is a Roman interpolation: the original Greek form was Iber (modern Ebro). It is the second largest river in the Iberian peninsula (930 km.) and from earliest times was a major route into the hinterland: livestock and grain sent downstream could be traded for Greek imports at Tyrichae. Palm products (normally dates and oil) are not generally considered a commodity of interior Iberia, and there may have been a confusion with imports from North Africa.[127]

Lines 504–19

The next section is from the Hiberus River to Tarraco. This is an extent of only about 50 km., an indication of how detailed the *Periplous* has become in these more familiar regions. The Sacred Mountain may be one of the promontories flanking the Ebro, and Oleum an alternative early name for that river, since there are no other significant streams on this coast. Mt. Sellus is probably the Serra del Boix, which rises to 941 m. just east of the river, or one of the Pyrenees foothills father north. Lebedontia would be the first coastal town north of the Ebro, but it was abandoned at an early date. Other deserted townsites were along this part of the coast: the name of Salauris may be preserved in the Salou district just south of Tarragona.

Callipolis (Kallipolis) was a major city with an extensive urban plan in a region of excellent fishing. This must be the later Tarraco (see line 519), since there is no other city in the region that could be described so expansively. Kallipolis presumably was the original name given by the Phokaian or Massalian settlers, but by no later than the second century BC the indigenous name of Tarraco, in hellenized form, had superseded the Greek one.[128] Avienus did not seem aware that Callipolis and Tarraco were at the same location.

Lines 519–29

Tarraco (Greek Tarrakon, modern Tarragona) was the major city of the region. Founded, probably as Kallipolis (lines 514–15), and a Phokaian or Massalian outpost, it has always been an important city and remains so today. There are extensive archaeological remains, mostly from the Roman period.[129] About 90 km. farther along the coast was the region of the Barcilones, around modern Barcelona. The town itself (Barcilo or Barcino) was not founded until perhaps the third century BC by the indigenous population at a location with a good harbor.[130]

The fierce Indigetes (the Indicetes of line 532) were the indigenous coastal population north of the region of Barcelona.[131] The Celebantic Ridge is one of the outliers of the Pyrenees, extending toward the coast north of Barcelona, perhaps the precipitous Cabo de Begur north of modern Palamós. By noting that it extended as far as "salty Thetis," or the Ocean, Avienus was making an allusion to the entire Pyrenees range. The city of Cypsela, probably a hellenization of an indigenous name, may have been the ancestor of the Phokaian trading center of Emporion, whose name survives as modern Ampurias or Empúries.[132]

Lines 530–43

Continuing to comment on Cypsela, Avienus recorded its particularly notable harbor, part of the large Golf de Roses, the most conspicuous maritime feature on this coast. The Indicetic shore is that of the Indicetes (line 532). The narrative moves into the coastal Pyrenees, with its increasingly rugged landscape and promontories projecting into the sea. Mt. Malodes is perhaps the uplands north of the Golf de Roses. The "twin summits" may refer to the extremities of the gulf; it created a calm bay and notable harbor well used from ancient to modern times.

Lines 544–65

The Toni lagoon and the adjacent Tononitian cliffs are features of the coast of the Pyrenees. There are several small streams in this region, but there is no certainty as to which might be the Anystus. The Ceretes ("Mountaineers") and their relatives the Ausoceretes lived in the upland hollows of the south side of the Pyrenees and were famous for their ham, still available today as the Serrano variety.[133] Although Avienus believed that they had assimilated into a general Hiberian ethnicity, they still retained their identity into the first century AD.

The Sordians were another mountain population in the heights of the Pyrenees, but would come down to the agricultural lowlands and coasts, perhaps as raiders but also to trade with the Massalians. The city of Pyrene cannot be specifically located, but must have been on the stretch of coast between Emporion and the summit of the mountains, or it may have been the original name of Emporion itself. It would have been the first Greek trading center in this region, known to Herodotus, the only other source to mention it by name, but his information was from a period when the topography of western Europe was so poorly known that he placed it near the source of the Istros (Danube).[134]

As a final note on the Iberian peninsula, and a detail of particular interest to mariners, it was recorded that from the Atlantic to Pyrene was a seven-day sail. This is a rough distance of about 1,200 km., easily possible in such a time period.

Lines 565–608

North of the Pyrenees was the extensive Cynetic Plain, the modern Roussillon, a name reflected in the ancient name for its major river, the Rhoscynus (modern Tech). The Sordicenian earth, if it refers to the Sordicene region of lines 552 and 558, would seem to be at the southern edge of the Cynetic Plain, but the name was also applied to the extensive local marshland, the modern Étang de Canet. The Sordus River is the modern Agly.

At this point two lines and part of a third are lost (lines 575–7). From here to the end of the extant *Ora Maritima* there are an increasing number of gaps, which Pisanus simply printed as blank spaces but preserved the line-by-line format. This probably indicates the deterioration of the manuscript near its end.

When the text resumes it is difficult to determine the context, since there are no toponyms for eight lines. Rather there is a navigator's description of the coast, curving along with its deep bays and numerous islands, which were a feature of the coast north of the Agly. Many of these have been lost in the bays and lagoons that are common today. The only islands mentioned are the Piplae, in the vicinity of Naro (modern Narbonne), which cannot be identified. The local population around Naro was the Elescyes, whose reputed fierceness may be because they were allied with the Carthaginians as early as the fifth century BC, a rare example of Carthaginian influence this far north.[135]

The Attagus (or Atax) River is the modern Aude, which flows across southern France and was part of an important route to the Atlantic. It empties into the Mediterranean just below Naro. The Helice swamp is one of the many lagoons in this region; because of reclamation and canalization the area has changed markedly from antiquity. Besara (or Baiterra) is modern Béziers, 30 km. northeast of Naro, on the Orobus (modern Orb) River. Archaeological evidence has demonstrated that the site was abandoned sometime betweeen 330/300 BC and 170/150 BC, and Avienus' explicit notice that Besara was no longer in existence (*stetisse*, line 591) is a good indication that his information came from a source produced during that period.[136] The Heledus is probably one of the affluents of the Orobus.[137] The desolate nature of the countryside has no chronological grounding, but was probably the situation before the rejuvenation after the Roman arrival in the second century BC.

The Thyrius River should be the next main stream, the modern Hérault, although this is not certain. Any elaboration is lacking because of another gap in the text, this time of three lines (596–8). Upon the resumption of the account Avienus was poetically describing the sea, with reference to the "quiet of the Alcyone." This is the European kingfisher (*Alceo atthis*), which was associated with the calm sea of late autumn (thus the term "halcyon").[138]

Any previous mention of Candidum is lost in the lacuna, and it must be located somewhere between the Thynius River and the island of Blasco. This is the modern Île de Brescou, near the mouth of the Hérault.[139] It was a region of extensive marshes and dunes, in the process of returning to its natural state due to the loss of population at the time of Avienus' source.

Lines 608–20

Mt. Setius (or Setion) is a conspicuous isolated hill (modern Mt. St-Clair) on the coast just west of the modern town of Sète, which preserves the ancient name. There are only small streams between the Hérault and the Rhone, one of which must be the Oranus. It is unlikely, however, that it ever was the dividing line between the Ligurians and Iberians, which traditionally was at the Rhone. The accepted ethnic boundary had moved to the Pyrenees by the third century BC.[140] Any relationship between Mt. Setius and the Taurus is peculiar, and the idea may be based on a regional toponym, probably reflected in the lagoon known today as the Bassin de Thau, southwest of Sète. Avienus has taken this local name and connected it linguistically with the Taurus Mountains of Asia Minor, the great range that was believed (at least in the eastern Mediterranean) to divide the inhabited world into a northern and southern half, based on the understanding that there was an almost unbroken chain from Asia Minor to India and the Himalayas.[141] Avienus may have believed that Mt. Setius was part of the Cévennes (the Cimenice of line 622) —whose southern end is about 75 km. to the north—and thus connected to the Alps, which stretched well to the east. There was also the ethnym Tauriskoi, which extended over a wide area of the eastern Alps (it survives in the modern Tauern region of Austria), which might also give credence to an association with the Taurus.

The three towns mentioned, Polygium, Mansa, and Naustalo, as well as the probable ethnym Hausicae (the text is unclear) must lie between Sète and the Rhone, but have not been located. Near the Rhone itself the topography has greatly changed due to siltation, channelization, and the movement of the river and its many mouths. Line 617 is defective, and the three following are missing.

Lines 621–30

The extant text resumes with the Classius River, the last stream before the Rhone. To the north is the Cimenice region (Greek Kemmenon, the modern Cévennes), a mountainous area west of the Rhone rising to 1702 m., one of the highest summits between the Alps and the Pyrenees.[142] The derivation of the name is quite plausible. The Rhodanus (Rhodanos, modern Rhone) flows east of the Cimenice, although Avienus was more poetic than accurate in suggesting that the Cimenice overhangs the sea, as it is quite a way inland. Setiena is the region of Mt. Setius (lines 608–9), which Avienus saw as the western extent of the Ligurian territory.

Lines 630–6

Line 630 introduces a long section (through line 699) on the Rhodanos and the Alps, in which the river has its source. Since it was an important part of the Massalian economy, such an excursis is not unexpected even if it violates the *periplous* format. This is perhaps the most purely geographical section of the *Ora Maritima*.

The Rhone, one of the major rivers of Europe, begins in central Switzerland, passes through Lake Geneva, and reaches the Mediterranean after a total route of 813 km. Aeschylus, who was remarkably astute geographically, seems to have been the first Greek to cite the river by name, in an unidentified play.[143] The context of Avienus' description of the river is clarified by his statement that it provided profit for its inhabitants: the river meant access to a broad region of western Europe and the resulting trade benefits. In addition, through short portages, the Massalians could reach the North Atlantic and beyond.

Avienus' disciple Probus was cited by name for the first time since line 51, and for the only time beyond the introduction in the extant poem. This may be because Avienus felt that the digression about the Rhone was of particular importance to the *Ora Maritima*, but it is possibly nothing more than the technique of dropping the dedicatee's name into the poem for no obvious reason, as Lucretius did with his various citations of C. Memmius.

Lines 637–61

The account of the Rhone begins at its source in the Alps. Although the mountains were well known in Avienus' day, due to Roman penetrations, they were still imperfectly understood by the early sources that he was using. The first definite citation of them in Greek literature was by Herodotus,[144] who believed that the toponym referred to a river, not as strange as it might seem, because it suggests that Massalians had already gone up the Rhone into the mountains. "Gallic" (from Gallia) is Avienus' Latin term for the primary ethnic group of northwest Europe, derived from the Greek "Galatai."[145] It is not certain whether he applied the ethnym to parts of the Alps, or more conventionally to regions along the lower Rhone, but if "Galatai" or a similar term appeared in Avienus' Massalian sources, it would be the earliest documentation of the ethnym.

The "tempestuous wind" would be the mistral, more applicable to the lower Rhone. The source of the river is at the Rhone Glacier north of Gletsch: the cave is beneath the glacier, from which the river emerges. The Column of the Sun, although considered a local river, is presumably one of the summits of the Alps; there are several examples of the toponym surviving today, usually as Sonnenhorn. Nevertheless, the Column of the Sun of the *Ora Maritima* is an allusion to the highest ranges of the Alps that parallel the river on its southern side, including peaks such as the Matterhorn and Mt. Blanc, which, at 4804 m., is the highest not only in the Alps but in all of Europe.

A digression on the path of the sun (lines 646–73) is attributed to the Epicureans, and thus was, in part, an insertion from the third century BC or later. The use of the second person, presumably addressed to Probus, shows that this material has been edited by Avienus. Yet the views expressed are not Epicurean, but actually an older Ionian view of the cosmos that describes how the sun circled the world but does not shine everywhere all the time. This idea is attributed to Anaximenes of Miletos, the third of the Ionian monists, active in the mid-sixth century BC.[146] The comments in the *Ora Maritima* were certainly stimulated by the effect of the high Alps on the visibility of the sun, since the mountains could block it for extended periods. It is possible that the material was in part from Avienus' Massalian sources, but its relevancy is marginal, and the attribution to the Epicureans removes it from any early context. Lines 658–61 are missing from the Pisanus edition.

Lines 662–73

When the text resumes, the context is still the path of the sun, using poetic expressions for the west (the Atlantic axis) and the east (the Achaemenid rising). The latter term, a peculiar phrase, refers to the Achaemenid dynasty of Persia, in power from 522–330 BC, and thus provides a chronological context for some of the material (notably before the time of Epicurus). The semi-mythical Hyperboreans ("Those Beyond the North") were an early feature of Greek understanding of the fringes of the world but were also based on actual knowledge of the peoples of northern Europe.[147] "Arctic" as a toponym ("near the Bear," referring to the constellation, but through metonomy meaning the far north) is not an early word, and may not have appeared in Avienus' original sources. The first known example in Greek is from the second century AD.[148]

Lines 674–98

Avienus returned to the course of the Rhone (from line 646). The Tylangi were the Tulangi encountered by Caesar, who lived near the Helvetii. The harshness of foreign ethnyms and toponyms was a regular Greco-Roman complaint.[149] The name of the Daliterni may survive with the Dala River, which joins the Rhone just below Leuk.[150] The Clahilci are otherwise unknown, but the name of the Lemenci survives with Lac Léman (Lake Geneva). Thus Avienus was describing the upper river, the portion in Switzerland, but there is no recognition of the lake itself.

The only toponym cited between the Lemenci and the area of the mouth of the river is Accion, obviously in a marshy region, and which is not located, although the name is known from the lower Rhone.[151] Arelatus is modern Arles, the last major town on the river, and which was probably a Phokaian foundation to control access to it, located just above its delta. The number of mouths of major rivers interested Greek geographers, perhaps due to the many mouths of the Nile. Reports on the Rhone varied from two to seven.[152] Today there are two, but the natural state of the region has long vanished.

Phileus (Phileas) of Athens was one of the sources mentioned at the beginning of the *Ora Maritima* (line 43). He probably lived in the second half of the fifth century BC.[153] Evidently the locals told him that the Rhone divided Europe from Libya, but this is so anomalous and so little is known about Phileus, including whether he actually made a trip to the lower Rhone, that the statement is impossible to contextualize. Continental theory was certainly new when Phileus wrote, having seemingly been established by Hekataios of Miletos, Aeschylus and Herodotus.[154] But there is no other evidence for this belief, which is geographically incomprehensible. Avienus was correct to be vigorously dismissive, and qualified Phileus' statement by noting that he merely "thought" (*putasse*) this to be the case. Line 698 is only partially extant but the context is clear, as it completes Avienus' excoration of Phileus.

The sailing time of two days and two nights, although without specified endpoints, must be from the Pyrenees, the previous location where a sailing figure was provided (lines 564–5). The terminus is equally uncertain, but somewhere around the mouth of the Rhone, perhaps even Arelate. The distance is about 250 km.

Lines 699–13

The *Ora Maritima* returns to the *periplous* format, which had not been used since line 630. The account moves southeast to Massalia, about 40 km. away from Arelate. The Nearchi are otherwise unknown. Bergine may be at modern Berre-l'Étang. The Salyes (or Sallusi) were the indigenous people between the Rhone and the Alps, whom the Phokaians and Massalians had to contend with, probably resulting in their reputation for fierceness.[155] The Mastrabalan lagoon is probably the modern Étang de Berre, south of the homonymous city, where the toponym Malestrou still survives near its entrance, although line 701 is corrupt and the exact form of the toponym is not certain.[156] The Cecylistrium Promontory is between that region and Massalia, possibly modern Cap Couronne.

The location of Massalia (Avienus' Massilia, modern Marseille) is particularly impressive, with an enclosed harbor that is not visible from the sea, and a permanent freshwater source at its head. The history of its foundation (around 600 BC) and government were described in detail by Strabo.[157] Clearly the situation impressed Avienus, or his source, and the visible remains today testify to the unique nature of the city. The final line of the extant text, as printed by Pisanus, tantalizingly suggests that Avienus may have provided some sort of updating of the material on Massalia, with more recent nomenclature, but the chronological context of this statement is lost.

With line 713, the extant text comes to an end. It is a suitably fitting terminal point for the surviving *Ora Maritima*, with its emphasis on the wonders of Massalia. At the end, Pisanus printed "The work of Rufius Festus Avienus ends," indicating that he had little idea that there was more to the treatise despite the end of the extant portion in the middle of a sentence. The description of Massalia may have marked the end of the first book of the *Ora Maritima*, and the text may have been truncated at this point long before it was printed.

Notes

1 Avienus, *Ora Maritima* (ed. Murphy) 100.
2 Because Avienus wrote in Latin, his form of personal names and toponyms often varies from the standard Greek orthography. In this discussion and the translation, Avienus' forms are used, with reference to the traditional Greek spellings as necessary.
3 *CIL* 6.537; John Matthews, "Continuity in a Roman Family: The Rufii Festi of Volsinii," *Historia* 16 (1967) 488–90.
4 A. H. M. Jones, "Collegiate Prefectures," *JRS* 54 (1964) 85–9; Luca Antonelli, *Il Periplo Nascosto* (Padua 1998) 153.
5 J. L. Lightfoot, *Dionysius Periegetes, Description of the Known World* (Oxford 2014).
6 Douglas Kidd, *Aratus*, Phaenomena (Cambridge 1997).
7 Polybios 34.15.7–9; Geminos 16.32.
8 Suetonius, *Divine Claudius* 42.
9 See, for example, the many instances at *Natural History* 6.178–80. Avienus occasionally used Greek endings (in Latin characters), probably for metrical reasons (see, for example, lines 91, 238, and 482).
10 Cicero, *Letters to Atticus* #24, 26.
11 Caesar, *Gallic War* 6.24; Pliny, *Natural History* 3.17; Roller, *Ancient Geography* 166–7.
12 Georgia L. Irby, "Tracing the *Orbis Terrarum* From Tingentera," in *New Directions in the Study of Ancient Geography* (ed. Duane W. Roller, University Park 2019) 104–7.
13 Johannes Engels, "Artemidoros of Ephesos and Strabo of Amaseia," in *Intorno al Papiro di Artemidoro* 2: *Geografia e Cartografia* (ed. C. Gallazi *et al.*, Rome 2012) 139–55.
14 Victor Buescu, *Cicéron: Les Aratea* (Hildesheim 1966); D. B. Gain, *The Aratus Ascribed to Germanicus Caesar* (London 1976).
15 Henry Mendell, "Euktemon," *EANS* 317; Paul T. Keyser, "Bakoris of Rhodes," *EANS* 188.
16 Rufus Festus Avienus, *Ora Maritima* (ed. Dietrich Stichtenoth, Darmstadt 1968) 10–11.
17 Mention of Octavian, the future Augustus, at line 279, is merely a chronological reference.
18 Pliny, *Natural History* 2.169.
19 Avienus (ed. Stichtenoth) 62.
20 Herodotus 4.152.

21 Herodotus 1.163.
22 Roller, *Through the Pillars* 10–12.
23 António Balboa Salgado, "Rufo Festo Avieno y su *Ora Maritima*: Consideraciones acerca de un sujecto y un objeto," *Gallaecia* 13 (1992) 369–98.
24 E. H. Bunbury, *A History of Ancient* Geography (second edition, London 1883) vol. 2, p. 685; Thomson, *History* 375; M. Cary and E. H. Warmington, *The Ancient Explorers* (Baltimore 1963) 44.
25 Carlo Santini, "Il prologo dell'*Ora Maritima* di Rufio Festo Avieno," in *Prefazioni, Prologhi, Proemi di Opere Technico-scientifico Latine* (ed. C. Santini and N. Scivoletto, Rome 1990–2) 937–47.
26 Herodotus 4.57 etc.
27 Sallust, *Histories* 3.82–99.
28 *FGrHist* #1.
29 *FGrHist* #4.
30 Paul T. Keyser, "Phileas of Athens," *EANS* 645.
31 *FGrHist* #709; Herodotus 4.44; Philip Kaplan, "Skulax of Karuanda," *EANS* 745–6.
32 Paul T. Keyser, "Pausimakhos of Samos," *EANS* 631.
33 *FGrHist* #5; Philip Kaplan, "Damastes of Sigeion," *EANS* 224–5.
34 Keyser, "Bakoris," *EANS* 188.
35 Mendell, "Euktemon," *EANS* 317.
36 Paul T. Keyser, "Kleon," *EANS* 481.
37 Plutarch, *On Exile* 13.604f.
38 Roller, *Eratosthenes' Geography* 263–7.
39 Homer, *Iliad* 18.483–9; Aristotle, *On the Heavens* 2.14.298a.
40 Herodotus 1.163, 4.152; Pausanias 6.19.3; Antonelli, *Il Periplo* 155; Manuel Álvarez Martí-Aguilar, "Arganthonius Gaditanus. La identificación de Gadir y Tarteso en la tradición antigua," *Klio* 89 (2007) 477–92.
41 *King Nikomedes Periodos* 189; Serena Bianchetti, "Avieno, Ora Mar. 80 ss.: Le Colonne d'Eracle e il vento del nord," *Sileno* 16 (1990) 241–6.
42 Pytheas T14; Strabo, *Geography* 4.4.1.
43 Barry Cunliffe, *Facing the Ocean: The Atlantic and Its Peoples* (Oxford 2001) 302–5.
44 Herodotus 3.115; Pliny, *Natural History* 7.197; Eduardo Ferrer-Albelda and Pedro Albuquerque, "El conocimiento del extremo Occidente en la Grecia arcaica: las Casitérides y la geografia de los recursos," in *La Ruta de las Estrímnides* (ed. Eduardo Ferrer Albelda, Alcalá de Henares 2019) 135–84.
45 Caesar, *Gallic War* 5.12.4; Healy, *Mining* 60–2.
46 Strabo, *Geography* 3.2.7, 15.2.12; Kitchell, *Animals* 197–9.
47 Caesar, *Civil War* 1.54; Casson, *Ships* 7; Cunliffe, *Facing* 66–8.
48 Casson, *Ships* 281–95.
49 Strabo, *Geography* 2.1.13 etc.; Philip Freeman, *Ireland and the Classical World* (Austin 2001) 28–33.
50 Pomponius Mela 3.53.
51 Pytheas T22 = Pliny, *Natural History* 4.102; Avienus (ed. Stichtenoth) 57.
52 1 Kings 10:22; López-Ruiz, "Tartessos" 255–80.
53 *King Nikomedes Periodos* 165.
54 Emmanuelle Meunier, "El estaño del noroeste ibérico desde la Edad del Bronce hasta la época romana. Per una primera síntesis," in *La Ruta de las Estrímnides* (ed. Eduardo Ferrer Albelda, Alcalá de Henares 2019) 279–320.
55 Pliny, *Natural History* 7.197.
56 Pliny, *Natural History* 2.169; Avienus, *Ora Maritima* 114–29, 380–9, 404–15; Geus, *Prosopographie* 157–9; Roller, *Through the Pillars* 27–9.

57 Sebastián Celestino and Carolina López-Ruiz, *Tartessos and the Phoenicians in Iberia* (Oxford 2016) 47–8; Roller, *Through the Pillars* 49–50.
58 *On Marvellous Things Heard* 136; Cary and Warmington, *Ancient Explorers* 46–7.
59 Pausanias 8.3.15; Apollodoros, *Bibliotheke* 3.8.2.
60 Strabo, *Geography* 4.6.2; Pliny, *Natural History* 37.52–3; Avienus (ed. Stichtenoth) 57–8.
61 Polybios 1.17.4, 7.9.5; Livy 29.2.2; Poseidonios F268.
62 Jean-Paul Morel, "Les Phocéens en Occident: certitudes et hypothèses," *PP* 21 (1966) 378–420; Celestino and López-Ruiz, *Tartessos* 88–9; Antonelli, *Il Periplo* 158–9.
63 Eratosthenes, *Geography* F157; Polybios 2.14; Pliny, *Natural History* 3.75.
64 Polybios 34.10.6; Diodoros 5.22; Strabo, *Geography* 3.2.11; Roller, *Through the Pillars* 68–9.
65 Barry Cunliffe, *The Extraordinary Voyage of Pytheas the Greek* (London 2001) 56–62.
66 *King Nikomedes Periodos* 405–6; Anaxagoras F37 Graham; Diogenes Laertios 2.8.
67 Jorge de Alarcão, "Novas perspectivas sobre os Lusitanos (e outros mundos)," *RPA* 4 (2001) 320–4.
68 Ptolemy, *Geographical Guide* 2.6.4.
69 Strabo, *Geography* 3.3.5; Pliny, *Natural History* 4.119.
70 Avienus (ed. Stichtenoth) 59.
71 Cunliffe, *Facing* 41–2; Maria Eugenia Aubet, "Mainake: The Legend and the New Archaeological Evidence," in *Mediterranean Urbanization 800–600 BC* (ed. Robin Osborne and Barry Cunliffe, Oxford 2006) 197–8.
72 Polybios 3.22; Pausanias 10.8.6, 10.18.7.
73 Dionysios Periegetes 338; Lightfoot, *Dionysius* 338–9.
74 José Cardim Ribeiro, "A *Ora maritima* de Avieno e a descrição da costa atlântica entre o Cabo da Roca e a foz do Sado. A propósito la localização de *Poetanion*," in *La Hispania Prerromana, Actas del VI Coloquio Sobre Lenguas y Culturas Prerromanas de la Península Ibérica* (ed. Francisco Vilar and José d'Encarnação, Salamanca 1996) 282–4.
75 Herodotus 2.33, 4.49; Strabo, *Geography* 3.1.4.
76 Stesichoros F S7; Strabo, *Geography* 3.2.11.
77 Arruda, "Phoenician Colonization" 123–8; for other suggestions, see Vasco Mantas, "A propósito de Ceuta: Algumas questões de geografia e epigrafia antigas," *Humanitas* 72 (2018) 88–95.
78 Homer, *Iliad* 8.13–16; Strabo, *Geography* 3.2.13.
79 Euripides, *Orestes* 265; Strabo, *Geography* 3.1.9.
80 Andrew T. Fear, "Odysseus and Spain," *Prometheus* 18 (1992) 22–6.
81 Homer, *Iliad* 3.368 etc.
82 Sallust, *Histories* 2.93; Livy F18; Strabo, *Geography* 3.3.7, 3.4.10; Avienus (ed. Murphy) 57.
83 Pausanias 6.19.3.
84 Herodotus 2.6; Strabo, *Geography* 16.2.5, 33; Antonelli, *Il Periplo* 166.
85 Hesiod, *Theogony* 287–94.
86 Velleius 1.2.3; Pomponius Mela 3.46; López-Ruiz, "Tarshish" 263–6; Celestino and López-Ruiz, *Tartessos* 180–1.
87 Strabo, *Geography* 3.5.3.
88 *FGrHist* #275; Roller, *World of Juba II*; Christian Gizewski, "Duoviri, Duumviri," *BNP* 4 (2004) 739–40.
89 Stephanos of Byzantion, "Ligystine."

90 Adolf Schulten, *Tartessos: Ein Beitrag zur ältesten Geschichte des Westens* (second edition, Hamburg 1950) 54.
91 Anakreon F31; Herodotus 1.163–5, 4.152.
92 Healy, *Mining* 179–80; Celestino and López-Ruiz, *Tartessos* 184–5.
93 Antonelli, *Il Periplo* 171–2; Javier Gómez Espelosín, "Iberia in the Greek Geographical Imagination," in *Colonial Encounters in Ancient Iberia* (ed. Michael Dietler and Carolina López-Ruiz, Chicago 2009) 289–92.
94 Stesichoros F S7; see also Eratosthenes, *Geography* F153; Strabo, *Geography* 3.2.11.
95 Celestino and López-Ruiz, *Tartessos* 233–4.
96 Avienus (ed. Murphy) 60.
97 Strabo, *Geography* 3.1.8; Pliny, *Natural History* 3.7, as Juno.
98 Luca Antonelli, "Aviénus et les colonnes d'Hercule," *MCV* 31 (1995) 77–83.
99 Homer, *Odyssey* 1.51–4; Pindar, *Olympian* 3.44 etc.
100 Cicero, *On the Manilian Law* 10 etc.; Caesar, *Gallic War* 1.1.
101 Plautus, *Asinaria* 11.
102 Homer, *Odyssey* 7.20.
103 *King Nikomedes Periodos* 143–6.
104 Avienus (ed. Murphy) 60–1.
105 Damastes (*FGrHist* #5) F2; Skylax (*FGrHist* #709) F8.
106 Strabo, *Geography* 2.5.18; Lightfoot, *Dionysius Periegetes* 273–6.
107 Herodotus 1.203; Aristotle, *Meteorologika* 2.1.354a; Eratosthenes, *Geography* F23; Roller, *Ancient Geography* 102–4.
108 Dionysios Periegetes 52–4.
109 J. B. Hall, "Notes on Avienius' *Ora Maritima*," *RFIC* 112 (1984) 194.
110 Pomponius Mela 2.94.
111 Hanno 1; Strabo, *Geography* 17.3.19.
112 Hekataios of Miletos F41; F. W. Walbank, *A Historical Commentary on Polybius* (Oxford 1957–79) vol. 1, p. 347.
113 Antonelli, *Il Periplo* 176.
114 *King Nikomedes Periodos* 146–7; Strabo, *Geography* 3.4.2; Aubet, "Mainake" 187–202.
115 Varro, *De Lingua Latina* 5.68; Horace, *Odes* 4.6.38; L. Richardson, Jr., *A New Topographical Dictionary of Ancient Rome* (Baltimore 1992) 238.
116 Strabo, *Geography* 3.5.1, 9.1.9.
117 Strabo, *Geography* 17.3.3.
118 Strabo, *Geography* 3.4.6; Antonelli, *Il Periplo* 179–80.
119 Pliny, *Natural History* 3.19.
120 A. J. Sanchez Perez and R. C. Alonso de la Cruz, "La ciudad fenicia de Herna (Guadamar del Segura, Alicante)," *RStudFen* 27 (1999) 127–31.
121 Diodoros 5.17.1; Strabo, *Geography* 3.5.1.
122 Strabo, *Geography* 3.4.6; María José Pena, "Avieno y las costas de Cataluña y Levante (II): *Hemeroskopeion-Dianium*," *Faventia* 15 (1993) 61–77; Roller, *Historical and Topographical Guide* 154.
123 See also *King Nikomedes Periodos* 200–1.
124 Hekataios of Miletos F46.
125 Strabo, *Geography* 3.4.6.
126 María José Pena, "Avieno y las costas de Cataluña y Levante I. *Tyrichae: *Tyrikaí, ¿'La Tiria'?*," *Faventia* 11 (1989) 13–15; Antonelli, *Il Periplo* 182–3.
127 Dalby, *Food* 113–14.
128 Polybios 3.95.5; Joaquim Icart Leonila, "Cal-lípolis fou Tàrraco," *Faventia* 15 (1993) 79–89.

129   J. Arce, "Tarraco," *PECS* 882–3.
130   J. Maluquer de Motes, "Barcino," *PECS* 142–3.
131   Pliny, *Natural History* 3.21.
132   Strabo, *Geography* 3.4.8; Avienus (ed. Murphy) 66.
133   Strabo, *Geography* 3.4.11; Pliny, *Natural History* 3.22; Martial 13.54.
134   Herodotus 2.33; John Hind, "Pyrene and the Date of the 'Massaliot Sailing Manual'," *RSA* 2 (1972) 39–52.
135   Herodotus 7.165.
136   Daniela Ugolini and Christian Olive, "Béziers et les côtes languedociennes dans l'*Ora Maritima* d'Avienus," *RAN* 20 (1987) 143–54.
137   Antonelli, *Il Periplo* 189.
138   W. Geoffrey Arnott, *Birds in the Ancient World From A to Z* (London 2007) 12–13.
139   Strabo, *Geography* 4.1.6.
140   Eratothenes, *Geography*, F133; Pliny, *Natural History* 37.32, quoting Aeschylus.
141   Roller, *Ancient Geography* 106.
142   Strabo, *Geography* 4.1.11.
143   Pliny, *Natural History* 37.32.
144   Herodotus 4.49.
145   Polybios 1.6.2.
146   Anaximenes F18 Graham.
147   Homeric *Hymn to Dionysos* 29; Pindar, *Pythian* 10.29–30; Herodotus 4.13, 32–6; Timothy P. Bridgman, *Hyperboreans: Myth and History in Celtic-Hellenic Contacts* (New York 2005).
148   Dionysios Periegetes 519, as *arktoios*.
149   Caesar, *Gallic War* 1.5; Strabo, *Geography* 13.2.6.
150   Avienus (ed. Murphy) 72.
151   Avienus (ed. Stichtenoth) 70.
152   Polybios 34.10.5; Ptolemy, *Geographical Guide* 2.10.2; Apollonios 4.634; see also Strabo, *Geography* 4.1.8.
153   Keyser, "Phileas of Athens," *EANS* 645.
154   Aeschylus, *Persians* 718; Herodotus 4.45; Roller, *Ancient Geography* 50–1.
155   Strabo, *Geography* 4.1.6.
156   J. Jannoray, "A propos d'Avienus 'Ora Maritima,' vers 701–2," *RA* 36 (1950) 77–83.
157   Strabo, *Geography* 4.1.4–5.

EPILOGUE

When Hanno and Himilco ventured outside the Pillars of Herakles, seeing new commercial opportunities for the Carthaginians, they essentially began the exploration of the Atlantic by Mediterranean peoples. At that time, around 500 BC, the world of the External Ocean was a dark and hostile environment that was frighteningly different from the calmer Mediterranean. Although the Mediterranean could itself be detrimential to shipping, in the Atlantic there were ferocious tides and masses of vegetation that could hold a ship fast. Most pernicious of all, in the Atlantic one could be lost forever, driven out of sight of land never to return: in the Mediterranean it was almost impossible to lose a view of the nearest coast. No one knew what lay across the Atlantic, if anything. Yet the Carthaginians, Phokaians, and Massalians headed beyond the Pillars of Herakles into this alien world, south to the tropics and north to the British Isles and beyond. The development of the *periplous* concept provided an increasing amount of information about the coasts: landmarks, river mouths, dangerous shoals, and hostile peoples. By the time the Romans arrived, in the second century BC, there was a body of received information, both oral and published, that meant a sail on the Atlantic, while always dangerous, was easier.

Hundreds of years later, when Avienus wrote, there was little left to explore, from the point of view of those living in the Mediterranean world. With traders and merchants leading the way, there were few areas of the Eastern Hemisphere that had not been reached by someone from the Mediterranean cultures. The Mediterranean system itself had been carefully plotted, and the rivers flowing into it explored nearly to their sources: the Danube, Dneiper, and Volga allowed access to central and northern Europe, and the Nile to central Africa. Expeditions had gone as far as the Arctic, Scandinavia, and the Baltic, and perfected the trade route from that sea to the Black Sea. To the east, there were routes to the Caspian and beyond, through Central Asia as far as China, and then south to India and east to the Malay Peninsula and Java. To the south, the upper reaches of the Nile had been visited, and traders had discovered the region of Zanzibar and the mysterious Selene Mountains, which as the Mountains of the Moon so excited Victorian explorers.

DOI: 10.4324/9781003030379-6

In fact, the only parts of the hemisphere that do not seem to have seen much penetration from the Mediterranean were remotest southern Africa, northeastern Siberia, and the extremities of southeastern Asia. It was even suggested that there might be unknown continents across the Atlantic and south of the equator. But all of this had happened by the mid-second century AD, over two centuries before Avienus wrote. In his day, exploration of unknown regions was a thing of the past, and Avienus' language demonstrates the mysterious nature of such endeavors, using phrases such as "obscure pronouncements," "the secret of things," or "deep matters." To Avienus, exploration was a thing of the "ancients," not something of his own era. His work was for a different environment, unlike Hanno and even the author of the *King Nikomedes Periodos*, who were very much involved in the world that they described. But nothing in the *Ora Maritima* reflected contemporary events, or even those of the previous two centuries. It was a time for reflection and exultation of heroic deeds, not new adventures.

Thus Hanno and Avienus mark a beginning and an end, with the Hellenistic *Periodos* between them. The *Ora Maritima* is a final statement on the era of ancient exploration; new worlds were not to be discovered until the Arab efforts of medieval times and the Europeans in the Renaissance.

BIBLIOGRAPHY

Alarcão, Jorge de. "Novas perspectivas sobre os Lusitanos (e outros mundos)," *RPA* 4 (2001) 293–349.

Álvarez Martí-Aguilar, Manuel. "Arganthonius Gaditanus. La identificación de Gadir y Tarteso en la tradición antigua," *Klio* 89 (2007) 477–492.

Amigues, Suzanne. "Végétaux étranges ou remarquables du Maroc antique d'après Strabon et Pline l'Ancien," *AntAfr* 38–9 (2002–3) 39–54.

Antonelli, Luca. "Aviénus et les colonnes d'Hercule," *MCV* 31 (1995) 77–83.

——. *Il Periplo Nascosto* (Padua 1998).

Arce, J. "Tarraco," *PECS* 882–883.

Arnaud, Pascal. "Ancient Mariners Between Experience and Common Sense Geography," in *Features of Common Sense Geography: Implicit Knowledge in Ancient Geographical Texts* (ed. Klaus Geus and Martin Thiering, Berlin 2014) 39–68.

Arnott, W. Geoffrey. *Birds in the Ancient World From A to Z* (London 2007).

Arruda, Ana Margarida. "Phoenician Colonization on the Atlantic Coast of the Iberian Peninsula," in *Colonial Encounters in Ancient Iberia* (ed. Michael Dietler and Carolina López-Ruiz, Chicago 2009) 113–130.

Aubet, María Eugenia. *The Phoenicians and the West* (tr. Mary Turton, second edition, Cambridge 2001).

——. "Mainake: The Legend and the New Archaeological Evidence," in *Mediterranean Urbanization 800–600 BC* (ed. Robin Osborne and Barry Cunliffe, Oxford 2006) 187–202.

Avienus, Rufus Festus. *Ora Maritima* (ed. J. P. Murphy, Chicago 1977).

——. *Ora Maritima* (ed. Dietrich Stichtenoth, Darmstadt 1968).

Balboa Salgado, António. "Rufo Festo Avieno y su Ora Maritima: Consideraciones acerca de un sujecto y un objeto," *Gallaecia* 13 (1992) 369–398.

Barruol, Guy and Michel Py. "Recherces récentes sur la ville antique d'Espeyran à Saint-Gilles-du-Gard," *RAN* 11 (1978) 19–100.

Belarte, Maria Carme. "Colonial Contacts and Protohistoric Indigenous Urbanism on the Mediterranean Coast of the Iberian Peninsula," in *Colonial Encounters in Ancient Iberia* (ed. Michael Dietler and Carolina López-Ruiz, Chicago 2009) 91–112.

Bianchetti, Serena. "Aspetti di geografia ephorea nei Giambi a Nicomede," *PP* 69 (2014) 751–780.

——. "Avieno, Ora Mar. 80 ss.: Le Colonne d'Eracle e il vento del nord," *Sileno* 16 (1990) 241–246.

Blomqvist, Jerker. *The Date and Origin of Hanno's Periplus* (Lund 1979).

——. "Reflections of Carthaginian Commercial Activity in Hanno's Periplus," *OrSue* 33–35 (1984–1986) 53–62.

Boardman, John. *The Greeks Overseas: Their Early Colonies and Trade* (fourth edition, New York 1999).

Bolton, J. D. P. *Aristeas of Proconnesus* (Oxford 1962).

Boshnakov, K. *Pseudo-Skymnos* (Stuttgart 2004).

Bradley, Guy. *Ancient Umbria* (Oxford 2000).

Braund, David. *Georgia in Antiquity* (Oxford 1994).

Bredow, Iris von. "Regnum Bosporanum," *BNP* 12 (2008) 443–450.

Bridgman, Timothy P. *Hyperboreans: Myth and History in Celtic-Hellenic Contacts* (New York 2005).

Buescu, Victor. *Cicéron: Les Aratea* (Hildesheim 1966).

Bunbury, E. H. *A History of Ancient Geography* (second edition, London 1883).

Burstein, Stanley. *Outpost of Hellenism: The Emergence of Heraclea on the Black Sea* (Berkeley 1976).

Camporeale, Giovannangelo. "The Etruscans and the Mediterranean," in *A Companion to the Etruscans* (ed. Sinclair Bell and Alexandra A. Carpino, Chichester 2015) 67–86.

Carcopino, Jérôme. *Le Maroc antique* (tenth edition, Paris 1948).

Cardim Ribeiro, José. "A Ora maritima de Avieno e a descrição da costa atlântica entre o Cabo da Roca e a foz do Sado. A propósito la localização de Poetanion," in *La Hispania Prerromana, Actas del VI Coloquio Sobre Lenguas y Culturas Prerromanas de la Península Ibérica* (ed. Francisco Vilar and José d'Encarnação, Salamanca 1996) 279–300.

Cary, M. and E. H. Warmington. *The Ancient Explorers* (Baltimore 1963).

Casson, Lionel. *Libraries in the Ancient World* (New Haven 2001).

——. *Ships and Seamanship in the Ancient World* (Princeton 1971).

Celestino, Sebastián and Carolina López-Ruiz, *Tartessos and the Phoenicians in Iberia* (Oxford 2016).

Clarke, Katherine. *Between Geography and History: Hellenistic Constructions of the Roman World* (Oxford 1999).

Cohen, Getzel M. *The Hellenistic Settlements in Europe, the Islands, and Asia Minor* (Berkeley 1996).

——. *The Hellenistic Settlements in Syria, the Red Sea Basin, and North Africa* (Berkeley 2006).

Coulson, W. D. E. "Taras," *PECS* 878–880.

Cunliffe, Barry. *The Extraordinary Voyage of Pytheas the Greek* (London 2001).

——. *Facing the Ocean: The Atlantic and its Peoples* (Oxford 2001).

Dalby, Andrew. *Food in the Ancient World From A to Z* (London 2003).

Demerliac, J.-G. and J. Meirat. *Hannon et l'empire punique* (Paris 1983).

Desanges, Jehan. "Lixos dans les sources littéraires grecques et latines," in *Actes du colloque de Larache, 8–11 novembre 1989* (Rome 1992) 1–6.

Dietler, Michael. "Colonial Encounters in Iberia and the Western Mediterranean: An Exploratory Framework," in *Colonial Encounters in Ancient Iberia* (ed. Michael Dietler and Carolina López-Ruiz, Chicago 2009) 3–48.

Diller, Aubrey. "The Ancient Measurements of the Earth," *Isis* 40 (1949) 6–9.

——. "The Authors Named Pausanias," *TAPA* 86 (1955) 268–279.

——. *The Tradition of the Minor Greek Geographers* (New York 1952).

Dridi, Hédi. "Early Carthage: From Its Foundation to the Battle of Himera (ca. 814–480 BCE)," in *The Oxford Handbook of the Phoenician and Punic Mediterranean* (ed. Carolina López-Ruiz and Brian R. Doak, Oxford 2019) 141–154.

Dunbabin, T. J. *The Western Greeks* (Oxford 1948).

Engels, Johannes. "Artemidoros of Ephesos and Strabo of Amaseia," in *Intorno al Papiro di Artemidoro 2: Geografia e Cartografia* (ed. C. Gallazi *et al.*, Rome 2012) 139–155.

Euzennat, M. "Jibila," *PECS* 426.

——. "Lixus," *PECS* 521.

——. "Pour une lecture marocaine du Périple d'Hannon," *BCTH* 12–14b (1976–1978) 243–246.

——. "Tingi," *PECS* 923.

Fear, Andrew T. "Odysseus and Spain," *Prometheus* 18 (1992) 19–26.

Ferrer-Albelda, Eduardo and Pedro Albuquerque. "El conocimiento del extremo Occidente en la Grecia arcaica: las Casitérides y la geografia de los recursos," in *La Ruta de las Estrímnides* (ed. Eduardo Ferrer-Albelda, Alcalá de Henares 2019) 135–184.

Fraser, P. M. *Ptolemaic Alexandria* (Oxford 1972).

Freeman, Philip. *Ireland and the Classical World* (Austin 2001).

Gärtner, Hans Armin. "Marcianus [1]," *BNP* 8 (2006) 304–305.

Gain, D. B. *The Aratus Ascribed to Germanicus Caesar* (London 1976).

Gelenius, S. *Arriani et Hannonis Periplus* (Basel 1533).

Geus, Klaus. *Prosopographie der literarisch bezeugten Karthager* (Leuven 1994).

Gisinger, F. "Skymnos [1]," *RE* 5 (second series, 1927) 661–687.

Gizewski, Christian. "Duoviri, Duumviri," *BNP* 4 (2004) 739–740.

Gómez Espelosín, Javier. "Iberia in the Greek Geographical Imagination," in *Colonial Encounters in Ancient Iberia* (ed. Michael Dietler and Carolina López-Ruiz, Chicago 2009) 281–297.

Graham, Daniel W. *The Texts of Early Greek Philosophy* (Cambridge 2010).

Hall, J. B. "Notes on Avienius' Ora Maritima," *RFIC* 112 (1984) 192–195.

Hammond, N. G. L. "The Peloponnese," *CAH* 3.1 (1982) 696–744.

Hanno. *Periplus* (ed. Al. N. Oikonomides and M. C. J. Miller, Chicago 1995).

Harden, D. B. "The Phoenicians on the West Coast of Africa," *Antiquity* 22 (1948) 141–150.

Head, Barclay V. *Historia Numorum* (Oxford 1911).

Healy, John F. *Mining and Metallurgy in the Greek and Roman World* (London 1978).

Heckel, Waldemar. *Who's Who in the Age of Alexander the Great* (Oxford 2006).

Hind, John. "Pyrene and the Date of the 'Massaliot Sailing Manual'," *RSA* 2 (1972) 39–52.

Hunter, Richard. "The Prologue of the Periodos to Nicomedes ('Pseudo-Scymnus')," *HG* 11 (2006) 123–140.

——. "Pseudo-Scymnus," in *Hellenistic Poetry: A Selection* (ed. David Sider, Ann Arbor 2016) 524–537.

Huss, Werner. "Carthage [IA]," *BNP* 2 (2003) 1130–1131.

Icart Leonila, Joaquim. "Cal-lípolis fou Tàrraco," *Faventia* 15 (1993) 79–89.

Irby, Georgia L. "Tracing the Orbis Terrarum from Tingentera," in *New Directions in the Study of Ancient Geography* (ed. Duane W. Roller, University Park 2019) 103–134.

Irby-Massie, Georgia L. "Semos of Delos," *EANS* 730–731.

Isserlin, B. S. J. *et al.* "The Canal of Xerxes: Summary of Investigations 1991–2001," *BSA* 98 (2003) 369–385.

Jannoray, J. "A propos d'Avienus 'Ora Maritima', vers 701–2," *RA* 36 (1950) 77–83.

Jones, A. H. M. "Collegiate Prefectures," *JRS* 54 (1964) 78–89.

Kaplan, Philip. "Damastes of Sigeion," *EANS* 224–225.

——. "Skulax of Karuanda," *EANS* 745–746.

Keyser, Paul T. "Bakoris of Rhodes," *EANS* 188.

——. "Kleon of Surakousai," *EANS* 481.

——. "Pausimakhos of Samos," *EANS* 631.

——. "Phileas of Athens," *EANS* 645.

Kidd, Douglas. *Aratus, Phaenomena* (Cambridge 1997).

Kindstrand, J. F. *Anacharsis: The Legend and the Apophthegmata* (Uppsala 1981).

Kitchell, Jr., Kenneth F. *Animals in the Ancient World From A to Z* (London 2014).

Korenjak, M. *Die Welt-Rundreise eines anonymen griechischen Autors ("Pseudo-Skymnos")* (Hildesheim 2003)

Lacroix, Francis. "Les langues," in *Histoire générale de l'Afrique noire* 1 (ed. Hubert Deschamps, Paris 1970) 273–290.

Lancel, Serge. *Carthage: A History* (tr. Antonia Nevill, Oxford 1995).

Lehoux, Daryn. "Diogenes of Babylon," *EANS* 253.

Lightfoot, J. L. *Dionysius Periegetes: Description of the Known World* (Oxford 2014).

Lightfoot, Jessica. "'Not Enduring the Wanderings of Odysseus': Poetry, Prose, and Patronage in Pseudo-Scymnus' Periodos to Nicomedes," *TAPA* 150 (2020) 379–413.

López-Ruiz, Carolina. "Phoenician Literature," in *The Oxford Handbook of the Phoenician and Punic Mediterranean* (ed. Carolina López-Ruiz and Brian R. Doak, Oxford 2019) 257–269.

——. "Tarshish and Tartessos Revisited: Textual Problems and Historical Implications," in *Colonial Encounters in Ancient Iberia* (ed. Michael Dietler and Carolina López-Ruiz, Chicago 2009) 255–280.

Maddoli, Gianfranco. "The Concept of 'Magna Graecia' and the Pythagoreans," in *Brill's Companion to Ancient Geography* (ed. Serena Bianchetti *et al.*, Leiden 2016) 43–57.

Magie, David. *Roman Rule in Asia Minor* (Princeton 1950).

Maluquer de Motes, J. "Barcino," *PECS* 142–143.

Manconi, D. "Sulcis," *PECS* 866.

Mantas, Vasco. "A propósito de Ceuta: Algumas questões de geografia e epigrafia antigas," *Humanitas* 72 (2018) 83–112.

Marcotte, Didier. *Les géographes grecques* 1 (Paris 2002).

Matthews, John. "Continuity in a Roman Family: The Rufii Festi of Volsinii," *Historia* 16 (1967) 484–509.

Mayor, Adrienne. *The Amazons* (Princeton 2014).

Mederos Martín, Alfredo. "La exploracíon del litoral atlántico norteafricano según el periplo de Hannón de Cartago," *Gerión* 33 (2015) 15–45.

Meineke, August. *Scymni Chii periegesis et Dionysii descriptio Graeciae* (Berlin 1846).

Mendell, Henry. "Euktemon of Athens," *EANS* 317.

Meunier, Emmanuelle. "El estaño del noroeste ibérico desde la Edad del Bronce hasta la época romana. Per una primera síntesis," in *La Ruta de las Estrímnides* (ed. Eduardo Ferrer Albelda, Alcalá de Henares 2019) 279–320.

Morel, Jean-Paul. "Les Phocéens en Occident: certitudes et hypothèses," *PP* 21 (1966) 378–420.

Müller, Carl. *Geographi graeci minores* (Paris 1855–1861).

Murray, G. J. "Trogodytica: The Red Sea Litoral in Ptolemaic Times," *GJ* 133 (1967) 24–33.

Palmer, H. R. "The Lixitae of Hanno," *JRAS* 27 (1927) 7–15.

Parke, H. W. *The Oracles of Apollo in Asia Minor* (London 1985).

Pearson, Lionel. *Early Ionian Historians* (Oxford 1939).

Pena, María José. "Avieno y las costas de Cataluña y Levante I. Tyrichae: *Tyrikaí,¿ 'La Tiria'?," *Faventia* 11 (1989) 9–21.

——. "Avieno y las costas de Cataluña y Levante (II): Hemeroskopeion-Dianium," *Faventia* 15 (1993) 61–77.

Pisanus, Victor. *Arati Phaenomena* (Venice 1488).

Ponsich, Michel. "Tanger antique," *ANRW* 2.10.2 (1982) 787–816.

Ramin, Jacques. *Le Périple d'Hannon* (*BAR Supplementary Series* 3, 1976).

Rebuffat, René. "Les nomades de Lixus," *BCTH* 18 (1982) 77–86.

——. "Les pentécontores d'Hannon," *Karthago* 23 (1995) 20–30.

Richardson, Jr., L. *A New Topographical Dictionary of Ancient Rome* (Baltimore 1992).

Roller, Duane W. *Ancient Geography: The Discovery of the World in Classical Greece and Rome* (London 2015).

——. *Empire of the Black Sea* (Oxford 2020).

——. *Eratosthenes' Geography* (Princeton 2010).

——. *A Historical and Topographical Guide to the Geography of Strabo* (Cambridge 2018).

——. *Scholarly Kings: The Writings of Juba II of Mauretania, Archelaos of Kappadokia, Herod the Great, and the Emperor Claudius* (Chicago 2004).

——. *Through the Pillars of Herakles: Greco-Roman Exploration of the Atlantic* (New York 2006).

——. "Timosthenes of Rhodes," in *New Directions in the Study of Ancient Geography* (ed. Duane W. Roller, University Park 2019) 56–79.

——. *The World of Juba II and Kleopatra Selene: Royal Scholarship on Rome's African Frontier* (London 2003).

Romer, F. E. *Pomponius Mela's Description of the World* (Ann Arbor 1998).

Roncaglia, Carolynn E. *Northern Italy in the Roman World* (Baltimore 2018).

Ryan, F. X. "Der sogennante Pseudo-Skymnos," *QUCC* 87 (2007) 137–143.

Sanchez Perez, A. J. and R. C. Alonso de la Cruz, "La ciudad fenicia de Herna (Guadamar del Segura, Alicante) ," *RStFen* 27 (1999) 127–131.

Santini, Carlo. "Il prologo dell'Ora Maritima di Rufio Festo Avieno," in *Prefazioni, Prologhi, Proemi di Opere Technico-scientifico Latine* (ed. C. Santini and N. Scivoletto, Rome 1990–1992) 937–947.

Savage, Thomas S. "Notice of the External Characteristics and Habits of Troglodyes Gorilla, a New Species of Orang From the Gaboon River," *Boston Journal of Natural History* 5 (1847) 417–443.

Scherf, Johannes. "Cecrops," *BNP* 3 (2003) 59–60.

Schulten, Adolf. *Iberische Landeskunde: Geographie des antiken Spanien* (Strasbourg 1955–1957).

———. *Tartessos: Ein Beitrag zur ältesten Geschichte des Westens* (second edition, Hamburg 1950).

Scullard, H. H. *The Elephant in the Greek and Roman World* (Cambridge 1974).

Segert, Stanislav. "Phoenician Background of Hanno's Periplus," *MUSJ* 45 (1969) 501–518.

Shipley, Graham. *Pseudo-Skylax's Periplous: The Circumnavigation of the Inhabited World* (Bristol 2011).

———. "Three Studies of 'Pseudo-Skymnos'," *CR* 57 (2007) 348–354.

Spadea, Giuseppina. "Terina e lo Pseudo-Scimno," *PP* 29 (1974) 81–83.

Stewart, George R. *Names on the Land: A Historical Account of Place-Naming in the United States* (New York 1945).

Stichtenoth, Dietrich. *Rufus Festus Avienus, Ora Maritima* (Darmstadt 1968).

Sullivan, Richard D. *Near Eastern Royalty and Rome, 100–30 BC* (Toronto 1990).

Sznycer, M. "Carthage et la civilisation punique," in *Rome et la conquête du monde méditerranéen* (ed. Claude Nicolet, Paris 1977–1978) 545–593.

Thomson, J. Oliver. *History of Ancient Geography* (Cambridge 1948).

Trofimova, Anna A. ed. *Greeks on the Black Sea* (Los Angeles 2007).

Tuchelt, K. "Didyma," *PECS* 272–273.

Tusa, V. "Panormos," *PECS* 671.

———. "Selinus," *PECS* 823–825.

Ugolini, Daniela and Christian Olive. "Béziers et les côtes languedociennes dans l'Ora Maritima d'Avienus," *RAN* 20 (1987) 143–154.

Vincenzo, Salvatore De. "Sicily," in *The Oxford Handbook of the Phoenician and Punic Mediterranean* (ed. Carolina López-Ruiz and Brian R. Doak, Oxford 2019) 537–552.

Walbank, F. W. *A Historical Commentary on Polybius* (Oxford 1957–1979).

West, Stephanie. "'The Most Marvellous of All Seas': The Greek Encounter with the Euxine," *G&R* 50 (2003) 151–167.

Xella, Paolo. "Religion," in *The Oxford Handbook of the Phoenician and Punic Mediterranean* (ed. Carolina López-Ruiz and Brian R. Doak, Oxford 2019) 273–292.

Yailenko, Valery. "Source Study Analysis of Pseudo-Scymnus' Data on The Pontic Cities' Foundation," *Pontica* 18–19 (2015–2016) 9–23.

PASSAGES CITED

Italicized numbers are citations in ancient texts; romanized numbers are pages in this volume.

Greek and Latin literary sources

Aeschylus
  Persians 718, 169n54; *865*, 117n317
  Prometheus Bound 349–52, 39n30
Agathemeros *2*, 112n67
Airs, Waters, and Places 13, 117n291
Aithiopis, Argument 4, 116n273
Anakreon *F31*, 168n91
Anaxagoras *F37*, 167n66
Anaximenes *F18*, 169n146
Andron of Teos (*FGrHist* #802) *F3*, 118n318
Antiochos of Syracuse (*FGrHist* #555) *F7*, 112n90
Apollodoros
  Bibliotheke
    Book 1: *7.2*, 115n209; *9.17*, 115n234
    Book 2: *5.8*, 116n242
    Book 3: *4.1*, 115n213; *5.4*, 114n55; *7.7*, 163n114; *8.2*, 167n59; *12.6*, 115n96
  F25, 111n27; *F53*, 111n27; *F170*, 117n370
Apollonios of Rhodes, *Argonautika*
  Book 2: *392*, 117n308; *964*, 117n393
  Book 4: *288–93*, 116n270; *309–12*, 116n272; *323–8*, 112n73; *569–71*, 114n151; *634*, 117n299
Appian, *Mithridateios 4–7*, 110n15; *67*, 117n299; *102*, 117n299
Aristophanes, *Acharnanians 75*, 115n98
Aristotle
  Meteorologika 1.9, 113n140; *1.13.350b*, 40n78; *1.13.351a*, 112n81; *2.1.354a*, 168n107; *2.5*, 111n42; *2.5.362b*, 6n17
  On the Heavens 2.13.294a, 6n11; *2.14.298a*, 166n39; *2.14.298b*, 6n15

DOI: 10.4324/9781003030379-7

Roman Antiquities
> Book 1: *11*, 112n97; *22*, 112n83; *27*, 112n87; *35.3*, 112n87; *61*, 116n245;
> *72*, 112n89

Dionysios Periegetes *26–67*, 154; *52–4*, 168n108; *338*, 167n73; *519*, 169n148

Diogenes Laertios
> Book 1: *101–5*, 117n286
> Book 2: *8*, 167n66
> Book 8: *3*, 113n127; *48*, 6n12
> Book 9: *20*, 111n19; *21*, 6n12

Empedokles of Akragas *F1*, 111n26

Ephorus (*FGrHist* #70) *F18b*, 114n183; *F30*, 112n66; *F30a*, 38n18, 117n282;
> *F34*, 115n232; *F43*, 117n311; *F44a*, 118n322; *F53*, 39n52; *F113*, 112n82;
> *F122a*, 114n170; *F128*, 112n60; *F129a*, 112n59; *F131b*, 112n69; *F136*,
> 113n102; *F137a*, 113n105; *F139*, 113n124; *F144*, 55, 144n70; *F145*, 56,
> 115n190; *F151*, 115n200; *F157*, 116n269; *F158*, 61, 117n283; *F159*, 61; *F160*,
> 61, 117n291; *F160a*, 117n293; *F161b*, 117n309; *F162*, 117n316; *T32*, 49

Eratosthenes
> *FGrHist* #241, *F1a*, 111n34
>
> *Geography F13*, 38n6, 116n238; *F23*, 168n107; *F25*, 6n13; *F30–1*,
> 111n35; *F133*, 169n40; *F146*, 54, 169n40; *F148*, 116n269; *F148–9*,
> 116n272; *F153*, 168n94; *F157*, 167n63

Euripides
> *Bakchai 677*, 50n58; *1330–9*, 114n155
> *Iphigeneia Among the Taurians 438*, 116n273
> *Orestes 265*, 167n79
> *Phoenician Women 666–75*, 115n273; *818–21*, 115n213
> *Trojan Women 884*, 41n93

Geminos *16.32*, 41n94, 165n7

Hanno, *Periplous 1* 6n6, 39n27, 168n111

Hekataios of Abdera (*FGrHist* #264) *F7–14*, 38n13; *F10*, 117n290; *F13*, 117n290

Hekataios of Miletos (*FGrHist* #1) *F41*, 112n168; *F46*, 168n24; *F62*, 112n87;
> *F84*, 113n26; *F86–8*, 113n133; *F91–2*, 113n32; *F170*, 116n270; *F291*,
> 117n302; *F355*, 40n55; *F357*, 39n53

Hellanikos of Lesbos (*FGrHist* #4) *F151*, 115n213

Herodotus
> Book 1: *6*, 155n267, 117n317; *15*, 116n267; *72*, 117n313; *75*, 117n317;
> *94*, 112n83; *96*, 39n49; *103–6*, 117n287; *145*, 113n119; *152*, 39n34;
> *163*, 112n76, 116n21, 40; *163–5*, 168n91; *166*, 38n19, 39n27; *167*,
> 112n99; *171*, 115n21; *203*, 168n207, 117n302
>
> Book 2: *6*, 167n84; *10*, 114n169; *32*, 39n44; *33*, 39n30, 167n75, 169n134;
> *34*, 63, 117n315, 118n318; *44*, 39n29; *68–70*, 40n80; *125*, 40n67; *154*,
> 39n49, 40n67
>
> Book 3: *71*, 40n81; *92–3*, 117n303; *94*, 117n311; *115*, 114n142, 166n44; *138*,
> 39n38

Book 2: *6.4*, 167n68; *10.2*, 169n152

Book 3: *5.7*, 116n277

Book 4: *6.9*, 41n94; *6.16*, 41n94; *6.33*, 40n75

Pytheas of Massalia *T14*, 166n42; *T22*, 166n51

Sallust, *Histories 2.93*, 167n82; *3.82–99*, 166n27

Scholia to Apollonios, *Argonautika 2.392*, 117n308

Seneca, *Natural Questions 3.14.1*, 6n11; *4a.2.22*, 11n16

Skylax of Karyanda (*FGrHist* #709) *F8*, 168n5; *T1*, 6n4

Sophokles

    Triptolemos F602, 39n27

    Women of Trachis 762, 40n58

Stephanos of Byzantion, *Ligystine*, 167n89; *Mentores*, 114n145

Stesichoros *F S7*, 167n76, 168n94

Strabo, *Geography*

    Book 1: *1.1*, 111n41; *2.28*, 38n18, 112n66; *3.1*, 116n238; *3.2*, 240n75; *3.21*, 116n267, 117n305

    Book 2: *1.13*, 166n49; *1.35*, 114n180; *1.41*, 116n274; *2.2*, 6n14; *3.4*, 41n101; *3.5*, 116n238; *5.15*, 112n65; *5.18*, 168n106; *5.20*, 112n87

    Book 3: *1.4*, 167n75; *1.8*, 168n97; *1.9*, 167n79; *2.7*, 112n62; *2.11*, 167n64, 167n76, 168n94; *2.13*, 167n78; *3.5*, 167n69; *3.7*, 167n82; *4.2*, 111n57, 168n114; *4.6*, 168n118, 122, 125; *4.8*, 112n77, 169n132; *4.10*, 167n82; *4.11*, 169n133; *4.18*, 117n293; *5.1*, 168n116, 121; *5.3*, 112n59; *5.4*, 112n59; *5.5*, 111n56; *5.6*, 112n71; *5.5–6*, 39n41

    Book 4: *1.4–5*, 112n76, 169n157; *1.6*, 169n139, 155; *1.8*, 169n152; *1.11*, 169n142; *4.1*, 112n72, 166n42; *6.2*, 167n60

    Book 5: *1.4*, 112n72, 114n141; *1.9*, 114n142; *2.1*, 112n83; *2.3–4*, 112n82; *4.3*, 112n90; *4.7*, 112n94; *4.11*, 112n95; *4.13*, 112n98

    Book 6: *1.1*, 112n99; *1.1–3*, 112n96; *1.4*, 112n97; *1.5*, 112n121, 122; *1.6*, 133n123; *1.8*, 113n124; *1.10*, 113n126; *1.12*, 112n127; *2.1*, 113n103; *2.2*, 113n105, 108; *2.3*, 113n111; *2.4*, 113n102, 110; *2.5*, 113n116; *2.6*, 113n104, 111; *2.10–11*, 113n100; *3.1*, 113n133; *3.1–3*, 113n130; *3.6*, 113n134; *3.9*, 114n152

    Book 7: *3.15*, 116n269; *3.17*, 116n276; *3.19*, 116n277; *4.4*, 117n281; *5.4*, 114n146; *5.5*, 114n148; *5.9*, 113n132; *6.1*, 116n258; *7.4*, 115n222, 117n281; *7.5*, 114n158; *7.7*, 114n152, 163; *7.8*, 114n146, 155, 115n222; *7.10–12*, 114n161; *F11*, 115n225, 228; *F14*, 115n226; *F16*, 115n231, 116n238, 240; *F18*, 116n242; *F20*, 116n245; *F21*, 116n248, 252

    Book 8: *3.33*, 114n172; *5.1*, 114n179; *5.4*, 114n188; *6.13*, 115n204; *6.16*, 115n196; *6.19*, 114n182; *6.20*, 114n178; *8.5*, 114n184

    Book 9: *1.9*, 168n114; *2.5*, 114n179; *3.1*, 114n188; *3.7*, 115n212; *3.42*, 117n299; *4.2*, 115n209; *4.8*, 114n178; *4.10–11*, 115n210; *4.11*, 115n218; *4.13*, 115n211; *5.5*, 115n219; *5.10*, 112n212; *5.16*, 115n207; *5.23*, 115n210

Biblical sources

Epigraphical sources

INDEX

Ethnyms are generally listed with toponyms. Not all modern toponyms are included. Spelling variants are shown in parentheses after the main entry.

For Product Safety Concerns and Information please contact our EU
representative GPSR@taylorandfrancis.com
Taylor & Francis Verlag GmbH, Kaufingerstraße 24, 80331 München, Germany

* 9 7 8 1 0 3 2 1 1 2 9 1 6 *